# The physics of elementary particles

## L. J. Tassie

Department of Theoretical Physics
Research School of Physical Sciences
Australian National University
Canberra

A Halsted Press Book

**John Wiley & Sons**
New York

LONGMAN GROUP LIMITED
London
*Associated companies, branches and representatives
throughout the world*

© Longman Group Limited 1973

First published 1973

Published in the U.S.A.
by Halsted Press, a Division
of John Wiley & Sons, Inc.
New York.

ISBN 0 470 84575-9
Library of Congress Catalog Card No.: 72–12447


*Printed in Great Britain by
William Clowes & Sons, Limited
London, Beccles and Colchester*

# Preface

This book is primarily intended for physics students who will not become particle physicists, and a deliberate attempt has been made to emphasize those parts of particle physics which are applications of principles shared by other branches of physics or which may be useful in other branches of physics. However, for the student who may wish to pursue the study of particle physics further, some suggestions for further reading are given, and it is hoped that this book is a suitable bridge to the advanced texts of particle physics.

This book has grown out of lecture courses given to third year undergraduate students and fourth (final honours) year students. The later chapters of the book contain material that has been presented in various seminars. A result of these diverse origins of the book is that the material becomes more difficult as the reader progresses further through the main text.

To enable the book to be used by readers with varied amounts of preparation, some material has been placed in appendices. The appendices, in part, consist of material that students could reasonably be expected to know, but frequently do not know. Other reference material has also been placed in the appendices.

The exercises, given at the end of each chapter, should be regarded as an essential part of the book, as some topics are dealt with more in the exercises than in the main part of the text. Answers to even-numbered exercises are given at the end of the book.

A few references are given and are listed at the end of each chapter. The references have been chosen mainly on the basis of possible usefulness to the student. An attempt has been made to keep the list of references small so that there is some chance of the student looking at some of them.

I would like to thank all my colleagues, including students, at the Australian National University who helped in many ways with the preparation of the book.

*November 1972*

# Acknowledgements

We are grateful to the following for permission to reproduce copyright material:

The Australian Institute of Physics for the article, 'Regge poles and particle physics' which appeared in *The Australian Physicist*, **6**, No. 10, pp. 149–154, October 1969, and was written by L. J. Tassie. Academic Press for Figure 47.1; American Institute of Physics for Figures 24.1a, 41.2, 41.3, 41.5, 41.7, 46.1, 59.1 from *Physical Review*, for Figures 34.1, 34.2, 39.1, 41.4, 42.2, 42.3, 43.1, 43.2, 43.3, 44.2, 45.1, 45.2, 45.3, 67.2, 67.3, 69.2 from *Physical Review Letters*, for Figures 45.5 and F2 from *Physics Today*, and for Figure 61.1 from *Reviews of Modern Physics*; Blackwell Scientific Publications Ltd for Figures F3, F5, F6, F7 from *Science Progress*; Institute of Physics, Sweden, for Figure 69.5; Israel Program for Scientific Translation, Jerusalem for Figure 41.1; MacMillan and Co. Ltd for Figures 23.1 and 23.2 from *Nature*; North-Holland Publishing Co. for Figures 45.6, 67.4, 69.1 and Tables 37.1 and 41.1 from *Physics Letters*; *Nuovo Cimento* for Figure 45.4; Princeton University Press for Figure 41.6; Stanford Linear Accelerator Center, California, for Figure 69.5.

# Contents

PREFACE                                                                    v

**1  Familiar particles**                                                  1
   1  Introduction                                          1
   2  Photons                                               1
   3  Electrons                                             2
   4  Protons                                               3
   5  Neutrons                                              4
   6  Conservation laws and invariance principles           5
     References                                   7
     Exercises                                    7

**2  More particles**                                                      8
   7  Antiparticles                                         8
   8  Feynman diagrams                                     11
   9  $\beta$ Decay and the neutrino                       17
  10  The origin of nuclear forces                             20
  11  Pions                                                    22
     References                                  23
     Exercises                                   24

**3  Properties of the pion**                                             26
  12  The spin of the $\pi^+$                                  26
  13  Parity                                                   26
  14  The parity of the $\pi^-$                                30
  15  The spin and parity of the $\pi^0$                       31
  16  Parity and absolute conservation laws                    32
     References                                  33
     Exercises                                   34

**4  Nucleons and pions**                                                 35
  17  Isospin                                                  35
  18  Charge independence of nuclear forces                    38
  19  Isospin of pions                                         39
     References                                  41
     Exercises                                   41

**5    Magnetic moments**                                                    43

    20   Nucleon magnetic moments                                        43
    21   Anomalous magnetic moments of electron and muon     45
         References                                                       48
         Exercise                                                         48


**6    Strange particles**                                                  49

    22   Summary of known particles until 1947                         49
    23   Strange particles                                                50
    24   Associated production and strangeness                        51
    25   $K$-mesons                                                       54
    26   Hyperons                                                         55
         References                                                       57
         Exercises                                                        57


**7    Non-conservation of parity**                                        59

    27   The $\theta$–$\tau$ puzzle                                      59
    28   Polarization of $\beta$ particles                              61
    29   The two-component neutrino                                     66
    30   Non-conservation of parity in $\Lambda^0$ decay               67
    31   Invariance under $P$, $C$ and $T$                             69
    32   $CP$ invariance                                                 70
    33   Classification of interactions                                 72
         References                                                       74
         Exercises                                                        75


**8    Leptons**                                                            76

    34   Two kinds of neutrinos                                          76
    35   The handedness of the muon neutrino                           78
    36   Conservation of leptons                                        79
    37   Universal conservation laws                                    81
         References                                                       83
         Exercises                                                        83


**9    Neutral $K$-mesons and non-conservation of $CP$**                  84

    38   Neutral $K$-mesons                                              84
    39   Non-conservation of $CP$                                        88
         References                                                       92
         Exercises                                                        92

**10   Resonances**                                                                   93

  40   Introduction                                                              93
  41   Resonances in pion–nucleon scattering                                      95
  42   Detection of resonance particles by energy–momen-
       tum correlations                                                         103
  43   More baryon resonances                                                   109
  44   The discovery of the $\Omega^-$                                          112
  45   Meson resonances with $S = 0$                                            114
  46   Meson resonances with $S = \pm 1$                                        123
  47   Resonances in various channels                                           124
  48   Nomenclature                                                             126
       References                                                               129
       Exercises                                                                131

**11   *SU* (3) multiplets of hadrons**                                               133

  49   Introduction                                                             133
  50   Group theory in physics                                                  136
  51   $SU(3)$ classification of baryons and mesons                             138
  52   The quark model                                                          144
  53   The quark model of mesons                                                146
  54   Properties of quarks                                                     147
  55   Baryons                                                                  149
  56   Mass splitting in the meson multiplets                                   151
  57   Mass splitting for baryons                                               154
  58   Derivation of Gell-Mann–Okubo mass formula for
       octet                                                                    155
       References                                                               160
       Exercise                                                                 161

**12   Regge poles**                                                                  162

  59   Regge poles                                                              162
  60   Exchange forces                                                          165
  61   Application to particle physics                                          167
  62   Complications                                                            170
       References                                                               170
       Exercise                                                                 171

**13   *SU* (6)**                                                                     172

  63   The quark model and $SU(6)$                                              172
  64   Ratio of magnetic moments of neutron and proton                         173
       References                                                               174

**14   Electromagnetic interactions**                                 175

   65   Introduction                                      175
   66   Form factors                                      176
   67   The form factors of the proton                    178
   68   The form factors of the neutron                   182
   69   Inelastic scattering                              183
   70   $e^+$–$e^-$ colliding beams                        191
       References                                 191

**15   Epilogue**                                                     193

    **Appendices**                                   194

A   SUMMARY OF SPECIAL RELATIVITY                                     194

   A1   Introduction                                     194
   A2   Four-vectors                                     195
   A3   Transformation between laboratory frame and
       centre-of-mass frame                       196
   A4   Time dilatation                                  203
       References                                 203

B   QUANTUM MECHANICS                                                 204

   B1   Introduction                                     204
   B2   States and operators                             205
   B3   Angular momentum                                 207
   B4   Addition of angular momenta                      209
       References                                 211

C   C1   Lifetime                                                     212
   C2   Cross-section                                    213
       References                                 213

D   PRINCIPLE OF DETAILED BALANCE                                     214

E   RESONANCE OF CLASSICAL OSCILLATOR                                 217

F   EXPERIMENTAL METHODS OF HIGH-ENERGY PHYSICS                       221

   F1   Introduction                                     221
   F2   Particle accelerators                            221
   F3   Intersecting storage rings                       222
   F4   Particle detectors                               225

|   | F5 | Bubble chambers | 226 |
|   | F6 | Spark chambers | 229 |
|   |    | References | 231 |
| G | LIST OF PARTICLES | | 233 |
| H | PHYSICAL CONSTANTS | | 241 |
| I | ANSWERS TO EVEN-NUMBERED EXERCISES | | 243 |

# Familiar particles

## 1 Introduction

From the study of atomic physics and low-energy nuclear physics, a great deal has been learned about certain elementary particles. We begin by briefly reviewing the properties of the familiar elementary particles of atomic physics.

According to the special theory of relativity, reviewed briefly in Appendix A, each particle obeys the energy–momentum relation

$$E^2 = c^2(p^2 + M^2c^2) \qquad (1.1)$$

where $M$ is the mass of the particle when at rest, $p$ is the momentum and $E$ is the total energy of the particle. For a particle at rest

$$E = Mc^2 \qquad (1.2)$$

## 2 Photons

(1) The work of Planck on black-body radiation showed that light of frequency $v$ occurs in quanta, called photons, each having energy

$$E = hv \qquad (2.1)$$

Equation (2.1) was also confirmed by the photoelectric effect. The relation between energy and momentum in electromagnetic radiation yields for the momentum of the photon

$$p = E/c = hv/c \qquad (2.2)$$

Equation (2.2) was confirmed experimentally in the study of the Compton scattering of a photon by a free electron.

From equation (2.2)

$$E^2 = c^2p^2 \qquad (2.3)$$

and so the photon has a rest mass of zero.

Other properties of photons learnt from atomic physics are:

(2) Photons can be created and destroyed in arbitrary numbers, as, for instance, in bremsstrahlung – when a charged particle is accelerated, as by hitting a target, photons are given off.

(3) The analysis of black-body radiation shows that photons obey Bose–Einstein statistics – they are bosons. There can be an arbitrary number of photons in a given state, and the wave function of a system of photons must be symmetric with respect to interchange of any two photons.

(4) The photon has spin 1. (More accurately, the square of the angular momentum of the photon has the value

$$2\hbar^2 \,=\, 1(1+1)\hbar^2$$

It is convenient to refer to a particle whose square of the angular momentum is

$$s(s+1)\hbar^2$$

as having spin $s$.)

There are two spin states for the photon, with $m_s = \pm 1$. $m_s \hbar$ is the $z$-component of the angular momentum, and the $z$-axis is taken along the direction of motion of the photon. These two spin states correspond to the two types of circularly polarized light. There is no spin state with $m_s = 0$.

The usual result from atomic spectroscopy of there being $2s+1$ spin states for spin $s$ holds only for a particle for which a frame of reference can be found in which the particle is at rest, and so holds only for particles with non-zero rest mass. No rest frame can be found for the photon which moves with the velocity of light $c$ in all reference frames.

## 3 Electrons

(1) The electron, symbol $e$, was the first elementary particle to be discovered. It has negative charge $-e$ (where $e = 1 \cdot 6022 \times 10^{-19}$ C) and mass

$$M_e \,=\, 9 \cdot 1096 \times 10^{-28} \text{ g}$$

Masses can be measured in the units of energy by using the rest energy $Mc^2$ in place of the mass $M$. In this way, the rest mass of a particle is frequently given in MeV (million electron volts) (see

Appendix H). For the electron,

$$M_e = 0.511 \text{ MeV}$$

(2) The study of atomic spectra showed that the electron has two spin states. The electron has non-zero rest mass, and the number of spin states must be $2s + 1$ where $s$ is the spin. Thus the electron has spin

$$s = \tfrac{1}{2}$$

(3) Electrons obey Fermi–Dirac statistics; they are fermions. The wave function of a system of electrons is antisymmetric with respect to the interchange of any two electrons; and so there can be at most one electron in a given state – the Pauli exclusion principle. The Pauli exclusion principle can be illustrated by considering a system of two (non-interacting) electrons, one in a state with wave function $\psi$ and the other in a state with wave function $\phi$. Then the total wave function $\Psi$, antisymmetric with respect to the interchange of electrons 1 and 2, is

$$\Psi(1, 2) = \psi(1)\phi(2) - \phi(1)\psi(2)$$

where $1$ and $2$ stand for all the coordinates (including spin) of electrons 1 and 2 respectively. We see that if $\phi = \psi$, then

$$\Psi = 0$$

– the two electrons cannot occupy the same state.

(4) Electric charge is conserved, and so electrons are not arbitrarily created or destroyed. The creation or destruction of an electron is always accompanied by the creation or destruction of some other particle or particles, as we shall see in more detail later.

## 4 Protons

(1) The proton, symbol $p$, which is the nucleus of the hydrogen atom, has charge $+e$ and mass

$$M_p = 938.3 \text{ MeV}$$

(2) The study of molecular hydrogen showed that the two protons in the hydrogen molecule could be arranged in two different ways. The spins of the two protons could be parallel as in orthohydrogen, or antiparallel as in parahydrogen. Each proton had two possible spin orientations relative to the spin of the other proton, and so, like the electron, the proton has spin $\tfrac{1}{2}$.

(3) In orthohydrogen, the wave function is symmetric with respect to interchange of the spins of the two protons since the two spins have the same direction, and experiment showed that the wave function was antisymmetric with respect to the interchange of the spatial coordinates of the two protons; so that the wave function is antisymmetric with respect to complete interchange of the two protons. In parahydrogen, the wave function is also antisymmetric with respect to complete interchange of the two protons, being antisymmetric with respect to interchange of the protons' spins, and symmetric with respect to interchange of their spatial coordinates.

So protons obey Fermi–Dirac statistics; they are fermions; the Pauli exclusion principle applies to protons – there can be at most one proton in any given state.

## 5 Neutrons

The neutron has symbol $n$ and mass $M_n = 939 \cdot 6$ MeV.

In 1930 Bothe and Becker discovered a very penetrating radiation given off when beryllium was bombarded with $\alpha$ particles; this penetrating radiation was thought to be $\gamma$ rays. In 1932 I. Joliot-Curie and J. F. Joliot-Curie found that this radiation knocked out protons from hydrogen-rich material, and they suggested that this was due to Compton scattering – i.e. that the protons were recoiling from scattering $\gamma$ rays. However, this explanation required the penetrating radiation to consist of extremely energetic $\gamma$ rays, but with no explanation of where such energy came from.

In 1932 Chadwick showed that the recoiling protons had been hit by neutral particles of approximately the same mass as the proton. He called these neutral particles neutrons. The reaction occurring when beryllium was bombarded with $\alpha$ particles was

$$_2\text{He}^4 + _4\text{Be}^9 \rightarrow _6\text{C}^{12} + _0n^1$$

The existence of the neutron was also needed to explain observations of molecular spectra, which showed, for instance, that the wave functions of nitrogen molecules were symmetric with respect to the interchange of the two $\text{N}^{14}$ nuclei, and consequently that the $\text{N}^{14}$ nuclei were bosons. This could not be understood if the $\text{N}^{14}$ nucleus was to be made up only of protons and electrons, as this would require 14 protons and 7 electrons, which constitutes an odd number of fermions. A system made up of an odd number of fermions is itself a fermion; for the interchange of two such systems can be carried out by interchanging their constituent fermions, and each interchange of two fermions changes the sign of the total wave function. In this way, it is also seen that a system made up of an even number of fermions will be a boson. Then the $N^{14}$ nucleus is a boson if made up of 7 protons and 7 neutrons, assuming that the neutron is a fermion.

From the study of nuclear physics, it is found that neutrons obey the Pauli exclusion principle and so are fermions, and also that the neutron has spin $\frac{1}{2}$.

It should be noted that particles with half-odd-integer $[(2n+1)/2]$ spins are fermions, and that particles with integral spin are bosons (Gamow, 1959). The proton and neutron have similar properties in many ways, and it is convenient to introduce the term 'nucleon' signifying either a neutron or a proton. This aspect of the neutron and proton will be discussed more fully in Sections 17 and 18.

The particles dealt with above, the photon, electron, proton and neutron, are sufficient for dealing with all of molecular and atomic physics. A few more particles are needed in the description of nuclear physics, a description that is still far from complete as our understanding of nuclear forces is very limited in comparison to our understanding of the forces in atomic and molecular physics. Finally, we shall encounter a seemingly never ending collection of particles in dealing with high-energy physics.

It might be argued that an understanding of the particles of high energy physics is an unnecessary luxury, and that we have sufficient understanding of our environment in terms of the familiar particles of atomic and molecular physics. However, we should remember that we do not yet understand nuclear forces, and so do not really understand why our surroundings are as they are, instead of being just clouds of hydrogen. In astronomy and cosmology, there are still many unsolved problems, such as the nature of quasars – the peculiar intense sources of energy which seem to be so small that it is difficult to understand how their great energy arises. It is very probable that the knowledge about elementary particles gained from high energy physics is relevant to the understanding and solution of problems in astronomy and cosmology. The science of elementary particles is an essential part of modern science with important consequences for other parts of science.

## 6 Conservation laws and invariance principles

In classical mechanics, conservation laws appear almost as an afterthought. For both the path of history, and the usual path trodden by the student, the equations of motion were first encountered, and the laws of conservation of momentum and conservation of mechanical energy were derived from the equations of motion. The conservation laws were then extended; for instance, the law of conservation of energy was extended to include chemical energy and electrical energy. However, in practical applications, the conservation laws are extremely useful, as they enable us to say something about complicated systems even when we do not know the detailed equations of motion of the systems. For instance, a full description of the collision of two automobiles would be very complicated, but we do know that momentum is conserved in such a collision. Similarly, in the case of collisions of particles, although we do not know the details of the interactions, energy, momentum and angular momentum are conserved in the collisions.

In high energy physics, where the equations of motion are as yet unknown, conservation laws are extremely important. The conservation laws of classical mechanics, conservation of energy, momentum and angular momentum, hold also in quantum mechanics. As we shall see later, there are also additional conservation laws in quantum mechanics.

Another aspect of the importance of conservation laws is that they are related to invariance principles or symmetry principles. An invariance principle states that the laws of physics remain unchanged (are invariant) for certain changes in circumstances. Or in the case of a particular system, an invariance property or symmetry property of the system is some operation that can be carried out on the system which does not alter the physics of the system.

As an example, the laws of physics are invariant under spatial translations. An experiment performed in London should yield the same answer as an experiment performed in New York. Also, the laws of physics are invariant under time translations; an experiment performed today should yield the same answer as the same experiment performed last year.

. In both classical mechanics (Landau and Lifshitz, 1969) and quantum mechanics (Feynman, 1965) invariance principles lead to conservation laws. For instance, invariance under spatial translations implies conservation of momentum; invariance under time translations implies conservation of energy; invariance under rotations implies conservation of angular momentum.

A large part of the study of elementary particles has been the search for further symmetries or approximate symmetries.

The treatment given here of the historical development of particle physics is necessarily very brief. Further information can be obtained from Boorse and Motz (1966), an anthology of important papers in the development of particle physics, together with commentary.

### References

BOORSE, H. A. and L. MOTZ (editors), *The World of the Atom*, 2 volumes, 1966. Basic Books, New York.

FEYNMAN, R. P., R. B. LEIGHTON and M. SANDS, *Quantum Mechanics*, Vol. III of *The Feynman Lectures on Physics*, 1965. Addison-Wesley. Reading, Mass. Chapter 17.

GAMOW, G., 'The exclusion principle', *Sci. Amer.* July 1959. (Also available as reprint 264, Freeman, San Francisco.)

LANDAU, L. D. and E. M. LIFSHITZ, *Mechanics*, Vol. I of *Course of Theoretical Physics*, 2nd edition, 1969. Pergamon, Oxford.

### Exercises

1  List evidence for and against the hypothesis that atomic nuclei are composed of protons and electrons.

2  Calculate the de Broglie wavelength of an electron with kinetic energy of (a) 10 eV, (b) equal to its rest energy, (c) 100 MeV.

3  Calculate the de Broglie wavelength of a proton with kinetic energy of (a) 10 eV, (b) equal to its rest energy, (c) 100 MeV.

4  At what kinetic energy does (a) a proton, (b) an electron, have a velocity one-half the velocity of light?

5  To probe the structure of an atomic nucleus, the wavelength of the probe particle must be smaller than the diameter of the nucleus. The radius of a nucleus of mass number $A$ is

$$r = 1 \cdot 2 \times 10^{-13} \times A^{\frac{1}{3}} \text{ cm}$$

so that a rough criterion is that the wavelength of the probe particle should be less than $10^{-12}$ cm. Calculate the energy of a (a) photon, (b) electron, (c) proton with wavelength of $10^{-12}$ cm.

# More particles

**2**

## 7 Antiparticles

In 1928 Dirac discovered a relativistic wave equation for a particle with spin $\frac{1}{2}$ – the Dirac equation. He showed that the Dirac equation provided a good description of the electron; for instance, the fine structure of the spectrum of the hydrogen atom calculated using the Dirac equation was in good agreement with experiment.

However there are difficulties because the Dirac equation has solutions with negative energy. For a free particle there are solutions with energy

$$E = \pm c(p^2 + M^2 c^2)^{\frac{1}{2}} \qquad (7.1)$$

Equation (7.1) also holds classically, but classically this is no problem because energy changes continuously and thus $E$ cannot change from positive to negative because of the gap between $+Mc^2$ and $-Mc^2$. In quantum mechanics, transitions can occur between states differing in energy by a finite amount; and so a transition from a positive energy state to a negative energy state is quite possible.

In order to avoid transitions of a positive energy electron to negative energy states, Dirac assumed that all the negative energy states are occupied, with one electron in each state in accordance with the Pauli exclusion principle. He also assumed that the occupied negative energy states cannot be observed. Then a positive energy electron cannot make a transition to a negative energy state, as all the negative energy states are occupied, and by the Pauli exclusion principle there can be at most only one electron in each state.

However, an electron occupying a negative energy state can undergo a transition to a positive energy state if sufficient energy is provided as, for instance, by an energetic photon. The unoccupied negative energy state now acts as a 'hole' in the 'sea' of negative energy electrons. This hole acts like a positively charged particle with the same mass as the electron and with positive energy. This particle is called the positron. Since the positron corresponds, in this description, to the absence of an electron, it is also called the antiparticle of the electron, or in short, the anti-electron. The symbols $e^-$ and $e^+$ are used for the electron and positron respectively.

In 1931 Anderson discovered the positron in a cosmic ray experiment.

After the absorption of a photon by an electron undergoing a transition from a negative energy state to a positive energy state, there is a positive energy electron and a positive energy positron, the latter corresponding to the unoccupied negative energy electron state, as is shown schematically in Fig. 7.1. Thus a photon has been transformed into an electron–positron pair. This process is called pair production, or pair creation. The photon must have at least enough energy to provide the rest energies of the electron and positron, namely

$$2Mc^2 = 1{\cdot}022 \text{ MeV}$$

so that electron–positron pair production has a threshold of 1·022 MeV.

Pair production cannot occur in free space, because the conversion of a photon into a pair cannot conserve both total energy and momentum (see Exercise 1). Some other particle must be present to take up whatever momentum and energy is required to conserve momentum and energy. The most commonly observed pair production is that occurring in the presence of an atomic nucleus, which can take up energy and momentum because of the interaction of its Coulomb field with either member of the pair. The nucleus is so much heavier than the electron, that the threshold for pair production can be calculated neglecting the energy transfer to the nucleus. Pair production by photons also occurs in the presence of another electron (see Exercise 2). Pairs are also produced in collisions of sufficiently energetic charged particles (see Exercise 3).

If one of the negative energy states of the electron is unoccupied, corresponding to the presence of a positron, then a positive energy electron can make a transition to the unoccupied negative energy state by emitting electromagnetic radiation in the form of photons. Thus a positive energy electron and a positron disappear and are replaced by photons. A transition of a positive energy electron to an unoccupied negative energy state represents the annihilation of an electron–positron pair (see Fig. 7.1). Because of conservation of energy and momentum, pair annihilation cannot occur with the emission of a single photon, and at least two photons must be emitted.

The Dirac equation describes a single electron. However, to interpret the negative energy solutions of the Dirac equation, an unobservable sea of an infinite number of negative energy electrons has been introduced, leading to a theory which is essentially a many-particle theory. So that the Dirac equation has only a limited

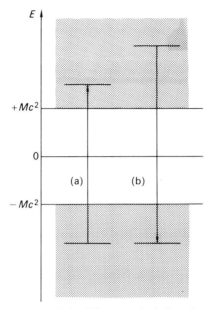

FIGURE 7.1    The two shaded regions represent the very closely spaced energy levels of the Dirac electron. (a) The creation of an electron–positron pair. (b) The annihilation of an electron–positron pair.

realm of applicability; it will yield accurate results only when the effect of the possibility of the creation and annihilation of particles is unimportant. The more general theory which describes arbitrary numbers of positrons and electrons in interaction with the electromagnetic field is known as quantum electrodynamics.

Since protons and neutrons have spin $\frac{1}{2}$ and are described by the Dirac equation, we expect to find antiprotons and antineutrons. In 1955, Segrè and his co-workers at Berkeley produced antiprotons using a beam of protons of 6·2 GeV kinetic energy from the Bevatron particle accelerator at the University of California (Segrè, 1956). Two years later, the antineutron was found.

An antiparticle is designated by the symbol for the corresponding particle with a bar placed over it, e.g. $\bar{p}$ antiproton, $\bar{n}$ antineutron.

## 8 Feynman diagrams

In his treatment of quantum electrodynamics in 1949, Feynman used the idea of representing elementary particles by space–time graphs, which are called Feynman diagrams. Such diagrams are useful in representing the processes that occur in the physics of elementary particles.

Particles move in four-dimensional space-time, but as it is much easier to draw two-dimensional diagrams, the three spatial coordinates are represented by one spatial coordinate $x$. Each particle corresponds to a line, called a world line, in the space–time graph. For instance, the world line of a free electron is a straight line, as in the Feynman diagram shown in Fig. 8.1, corresponding to the motion of the electron with uniform velocity. If we view this Feynman diagram through a horizontal slot, we see a one-dimensional space at a particular time. Let such a slot be called a time slot. The point where the world line of the electron intersects the time slot represents the position of the electron. If the slot is moved uniformly up the paper, we see the position of the electron changing uniformly with the time.

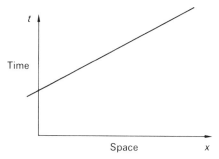

FIGURE 8.1   World line of a free particle.

Now consider the creation of an electron–positron pair, followed in time by the annihilation of the positron with another electron as shown in Fig. 8.2. Viewing the diagram through a moving time slot, we first see a single electron. Later, at $A$, a positron and an additional electron are created. At $B$ the positron and the initial electron annihilate each other. Finally the other electron continues to move with uniform velocity.

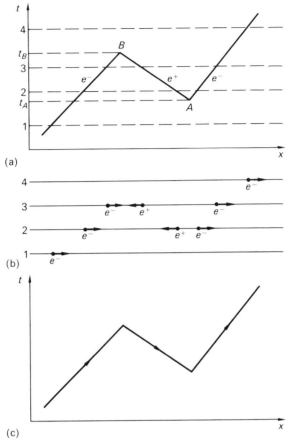

FIGURE 8.2   (a) Diagram for production of an electron–positron pair at $A$ followed by annihilation of the positron with another electron at $B$. (b) The same events as viewed through successive positions of a time slot, 1, 2, 3, 4, as shown in (a). (c) The final Feynman diagram for the process, as one continuous zig-zag line.

Before $t_A$, there is one world line; between $t_A$ and $t_B$ there are three world lines; and after $t_B$ there is again only one world line. However, there is only one continuous zig-zag line, although the positron part of this continuous line is directed backwards in time. Feynman pointed out the analogy of a pilot flying low over one road suddenly seeing three roads and, only when he sees two of them come together and disappear, realizing he has flown over a long S-bend in a single road.

To emphasize the singleness of the zig-zag world line, arrows are placed on each segment. The arrow on an electron segment points forwards in time; the arrow on a positron segment points backwards in time.

Similarly, world lines for other fermions can be drawn in Feynman diagrams, with an arrow on a particle segment pointing forwards in time, and an arrow on an antiparticle pointing backwards in time. Fermion lines have no beginning or end – fermions cannot be created or destroyed arbitrarily, but only as fermion–antifermion pairs.

We represent a photon on a Feynman diagram by a wavy line. The emission of a photon by an electron is shown in Fig. 8.3a. The diagrams for photon absorption, pair annihilation and pair production are shown in Figs. 8.3b, c, d respectively. We see that these different physical processes are represented by similar diagrams. For instance, the diagram for pair production can be obtained by twisting around the pieces of the diagram for photon emission.

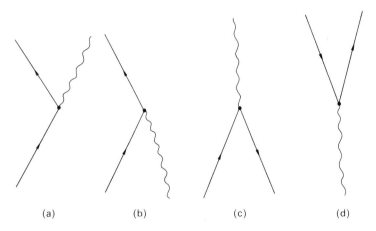

(a)                    (b)                    (c)                    (d)

FIGURE 8.3    (a) Photon emission. (b) Photon absorption. (c) Pair annihilation. (d) Pair creation.

The processes shown in Fig. 8.3 cannot conserve energy and momentum by themselves, and can only occur in the presence of other particles. Such processes are called virtual processes. A large diagram built up of diagrams for virtual processes can occur, provided conservation of energy and momentum is satisfied by the whole diagram. For instance, Fig. 8.4 shows a Feynman diagram for pair production in the Coulomb field of a proton. Each vertex in Fig. 8.4 is similar to one of the diagrams in Fig. 8.3, and consists of one photon line ending on a fermion line.

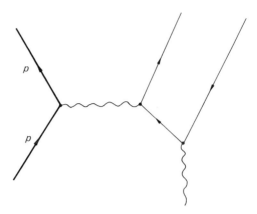

FIGURE 8.4    Pair production in the Coulomb field of a proton.

Because of the Heisenberg uncertainty principle in quantum mechanics, a particle cannot have a definite position in space-time and a definite energy and momentum. The more localized the particle is in space-time, the larger the uncertainty in its energy and momentum. So that, virtual processes which do not conserve energy and momentum can occur over very small intervals in space and time by virtue of the Heisenberg uncertainty principle, provided they are followed by processes which ensure conservation of energy and momentum for the whole process.

Another example, shown in Fig. 8.5, is the scattering of an electron by a proton.

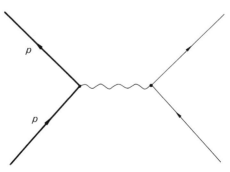

FIGURE 8.5    The scattering of an electron by a proton.

It should be pointed out that the exact positions of lines in a Feynman diagram are unimportant. It is only the topology of the diagram that is important. This is because the motion of particles must be described by quantum mechanics, in which particles do not have definite paths in space-time.

Feynman gave the rules for writing down the quantum mechanical probability amplitude corresponding to any Feynman diagram. In general, a process can be represented by many different Feynman diagrams, as is shown for electron–proton scattering in Fig. 8.6, and the probability amplitude is the sum of the probability amplitudes for each diagram. The probability is given by

$$\text{probability} = |\text{probability amplitude}|^2$$

The contribution to the amplitude from a diagram with $n$ vertices (each vertex being of the type shown in Fig. 8.3) contains a factor $(e/\sqrt{\hbar c})^n$. Since $e^2/\hbar c \approx 1/137$ is small, we expect the contribution from the lowest order diagram to predominate.

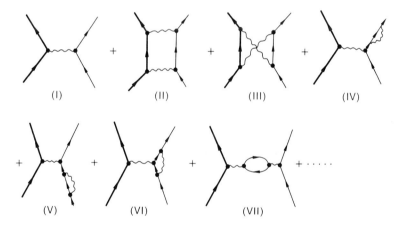

FIGURE 8.6   Some of the Feynman diagrams for the scattering of an electron by a proton.

However, using the Feynman rules, it is found that diagrams (iv), (v), (vi) and (vii) of Fig. 8.6 yield infinite contributions to the probability amplitude. These infinities are due to an inconsistent description of the electron and the photon. Higher order diagrams can be drawn for a free electron, as shown in Fig. 8.7, so that the electron now consists of its 'bare' self together with all its virtual interactions with the electromagnetic field, corresponding to the electron emitting and re-absorbing virtual photons. Similarly, higher order diagrams can be drawn for a free photon, as shown in Fig. 8.8. Thus the description of the electron and the photon is inconsistent with our starting point that the electron and photon are each represented by a single line.

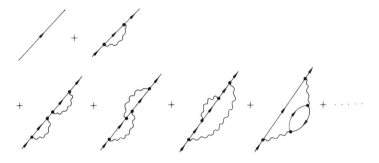

FIGURE 8.7    Feynman diagrams for a free electron.

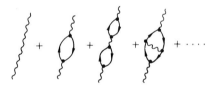

FIGURE 8.8    Feynman diagrams for a free photon.

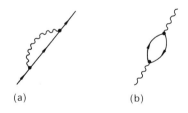

(a)                    (b)

FIGURE 8.9.

The theory can be corrected so that the free electron and free photon are correctly described (renormalization theory), the infinities from Feynman diagrams are removed, and all probabilities can be calculated to arbitrary accuracy. Then, in drawing Feynman diagrams, pieces like Fig. 8.9a should be omitted since this piece is inconsistent with the description of the electron, and pieces like

Fig. 8.9b should be omitted since this piece is inconsistent with the description of the photon.

The lowest order Feynman diagrams for electron–electron scattering[†] (Møller scattering) and electron–positron scattering (Bhabha scattering) are shown in Fig. 8.10, and other processes are considered in Exercises 4 and 7.

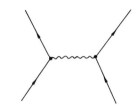

Electron - electron scattering

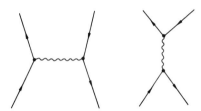

Electron - positron scattering

FIGURE 8.10.

## 9 β Decay and the neutrino

Let us denote the mass of a nucleus with mass number $A$ and atomic number $Z$ by $M_{A,Z}$. Then this nucleus is unstable with respect to $\beta$ decay, by the emission of an electron or a positron, if

$$M_{A,Z} > M_{A,Z \pm 1} + M_e \tag{9.1}$$

where $M_e$ is the mass of the electron.

Note that

$$M_n > M_p + M_e \tag{9.2}$$

and so the neutron is unstable with respect to $\beta$ decay.

---

† It should be noted that in determining the probability of electron–electron scattering from the probability amplitude, as given, for instance, from the Feynman diagram, Fig. 8.10a, the effect of the indistinguishability of the two electrons must be included. See Feynman (1965), Chapters 3 and 4.

When the $\beta$ decay of nuclei was investigated experimentally it appeared that neither energy nor spin were conserved. In order to rescue the conservation laws, Pauli postulated that another particle, the neutrino (symbol $v$), which has zero rest mass and spin $\frac{1}{2}$, is produced during $\beta$ decay. Being a spin $\frac{1}{2}$ particle, the neutrino also has an antiparticle, the antineutrino ($\bar{v}$).

The decay of the neutron is

$$n \to p + e^- + \bar{v} \tag{9.3}$$

An example of positron decay is

$$_{15}P^{30} \to {}_{14}Si^{30} + e^+ + v \tag{9.4}$$

which can be regarded as the decay of a proton inside the nucleus

$$p \to n + e^+ + v \tag{9.5}$$

Associating the antineutrino with electron emission, and the neutrino with positron emission, ensures the conservation of fermions, or more precisely the conservation of the difference in the number of fermions and the number of antifermions, in equations (9.3) and (9.5).

Regarding the neutron and proton as two different quantum states of the same particle, we can draw Feynman diagrams for the decays (9.3) and (9.5) as shown in Fig. 9.1. Note that four fermion lines meet at a vertex, two lines having arrows directed into the vertex, and two lines having arrows directed out from the vertex.

Neutrinos are also produced in $K$ capture, in which a nucleus decays after absorbing an electron from its $K$-shell. An example is

$$_{18}A^{37} + e^-_K \to {}_{17}Cl^{37} + v \tag{9.6}$$

This can be regarded as capture of an electron by a proton inside the nucleus,

$$p + e^- \to n + v \tag{9.7}$$

as depicted by the Feynman diagram, Fig. 9.1c. Equation (9.7) can be obtained from equation (9.5) using the rule that a particle on one side of the equation can be replaced by the corresponding antiparticle on the other side of the equation – an outgoing antiparticle is equivalent to an ingoing particle. Comparing Feynman diagrams, Figs. 9.1b and c, this rule corresponds to the convention for placing arrows on fermion lines.

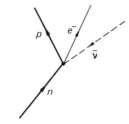

(a)  $n \rightarrow p + e^- + \bar{\nu}$

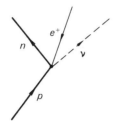

(b)  $p \rightarrow n + e^+ + \nu$

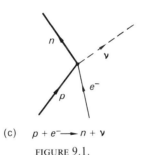

(c)  $p + e^- \rightarrow n + \nu$

FIGURE 9.1.

The existence of the neutrino was confirmed by experiments on the recoil of a nucleus emitting a neutrino and by the observation of inverse $\beta$ decay.

The simplest recoil experiments are those involving $K$ capture, since if the original atom is at rest, the final momentum of the recoiling ion must be equal and opposite to the momentum of the emitted neutrino. For instance, in the decay of $A^{37}$, equation (9.6), the energy of the emitted neutrino is determined from the $A^{37}$–$Cl^{37}$ mass difference to be $816 \pm 4$ keV. The $Cl^{37}$ ion should recoil with an energy of $9 \cdot 67 \pm 0 \cdot 08$ eV. Several measurements of this recoil energy have been made. Snell and Pleasonton (1955) obtained $9.63 \pm 0.06$ eV for the measured recoil energy, in good agreement with the predicted value, confirming that a single neutrino is emitted in the decay of $A^{37}$. See Allen (1958) Chapter 3.

The neutrino was first detected by Reines, Cowan and co-workers, using the reaction

$$\bar{v} + p \rightarrow e^+ + n \tag{9.8}$$

which may be considered as the inverse of the $\beta$ decay

$$n \rightarrow p + e^- + \bar{v} \tag{9.9}$$

A nuclear reactor provided an intense flux of antineutrinos, due to the $\beta$ decay of fission fragments. The antineutrinos were detected using a large liquid scintillator containing cadmium in addition to hydrogen. Then after the absorption of an antineutrino by a proton according to equation (9.8), the emitted positron promptly anni-hilates with an electron yielding two $\gamma$ rays which give rise to pulses of scintillation. The neutron first slows down by collisions, and is then captured by a cadmium nucleus which then emits one or more $\gamma$ rays which also cause a pulse of scintillation several microseconds after the prompt pulse due to positron annihilation. The light of the scintillation pulses is detected by arrays of photomultiplier tubes.

In the first experiment, the detector was a cylindrical volume of scintillator, with a diameter of 75 cm and a height of 75 cm, viewed by 90 photomultiplier tubes. In the second experiment, the detector consisted of a multiple-layer sandwich of three scintillation counters and two target tanks. The target tanks contained a water solution of cadmium chloride. The scintillation counters were each two feet thick, and the target tanks each three inches thick. Further details are given in Allen (1958), Chapter 7.

## 10 The origin of nuclear forces

According to the description by Feynman diagrams of Section 8, the Coulomb force between charged particles is due to the exchange of photons, as shown in Fig. 10.1a. The Coulomb force is a long range force, and the photon has zero rest mass.

The forces between nucleons have short range. In 1935 Yukawa suggested that these short range nuclear forces were due to the exchange between nucleons of particles of finite rest mass $M$ called mesons. This situation is depicted by the Feynman diagram, Fig. 10.1b. In the emission of a meson of rest mass $M$, energy will not be conserved by an amount $\Delta E = Mc^2$. By the Heisenberg uncertainty principle, $\Delta E \, \Delta t \gtrsim \hbar$, the exchanged meson could be emitted for a time $t$, where

$$t \simeq \hbar/\Delta E = \hbar/Mc^2$$

In this time the exchanged meson could travel at most a distance

$$R = ct \simeq \hbar/Mc$$

So the range of the nuclear force would be approximately $\hbar/Mc$.

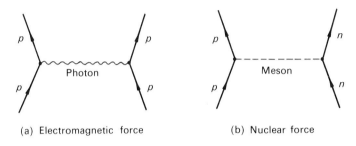

(a) Electromagnetic force  (b) Nuclear force

FIGURE 10.1.

Experimentally, the nuclear force is found to have a range of $R \approx 10^{-13}$ cm, yielding the estimate for the mass of the meson as

$$M \simeq \frac{\hbar}{Rc} \simeq 0{\cdot}3 \times 10^{-24} \text{ g}$$

$$\simeq 300 \, M_e$$

Thus in 1935 Yukawa predicted the existence of mesons which should interact strongly with nucleons as they gave rise to the strong nuclear forces, and with a mass $\sim 300 \, M_e$, intermediate between the mass of the electron and the mass of the nucleon.

In 1937 Anderson observed a particle of about the predicted mass in a cloud chamber experiment with cosmic rays. These particles were investigated during the following ten years, but, as their inter-action with nucleons was exceedingly weak, they could not be the mesons predicted by Yukawa.

This puzzle was solved by Lattes, Powell and Occhialini, who discovered there were two kinds of particles, called $\mu$-mesons and $\pi$-mesons. The $\pi$-meson interacts strongly with nucleons, but has a very short lifetime and decays to the $\mu$-meson, which was the particle previously discovered by Anderson. The $\mu$-meson has a longer lifetime and does not interact strongly with other particles. The existence of the $\mu$-meson (or muon, as it is also called) apparently has no relevance to nuclear forces. Indeed, the existence of the muon is still a puzzle to physicists, as its existence seems to have no relevance to any other physical phenomenon.

The $\pi$-meson (or pion, as it is also called) is the particle predicted by Yukawa, and the exchange of pions between nucleons contributes to nuclear forces. However, nuclear forces are not due completely to pions. Other mesons have been discovered, and will be discussed later. However, the masses of these other mesons are greater than the pion mass, and so their contribution to the nuclear force has a smaller range than the contribution from the pion. There is, as yet, no complete theory of nuclear forces. When nucleons are not too close, the nucleon–nucleon force is due to exchange of a single pion, as described by Fig. 10.1b, and the tail of the nucleon–nucleon interaction is adequately explained in this way. At closer distances, a satisfactory explanation of nuclear forces is still awaited.

Remembering that, according to the rules of Feynman diagrams, fermion lines cannot have ends, but only boson lines have ends, and noting that in diagram 10.1b, the pion line has ends, we conclude that if the exchange of pions contributes to the nuclear force, the pion must be a boson, and that pions can be produced or destroyed in arbitrary numbers.

Pions were first produced in the laboratory by Gardner and Lattes in 1948 using 380 MeV $\alpha$ particles from the 184 inch synchrocyclotron at the University of California.

## 11  Pions

The pion exists in three charge states, $\pi^+$, $\pi^0$, $\pi^-$. $\pi^+$ and $\pi^-$ have the same mass, 139·6 MeV, and the same lifetime $\tau = 2\cdot60 \times 10^{-8}$s (see Appendix C) and decay almost 100 per cent of the time by the processes

$$\pi^+ \rightarrow \mu^+ + \nu_\mu$$
$$\pi^- \rightarrow \mu^- + \bar{\nu}_\mu \tag{11.1}$$

although they also decay by other modes, such as in the fraction $1\cdot2 \times 10^{-4}$ by

$$\pi^+ \rightarrow e^+ + \nu_e$$
$$\pi^- \rightarrow e^- + \bar{\nu}_e \tag{11.2}$$

The fraction of decays into a particular mode of decay is called the branching ratio. The charged pions also decay by

$$\pi^+ \rightarrow \mu^+ + \nu_\mu + \gamma$$
$$\pi^- \rightarrow \mu^- + \bar{\nu}_\mu + \gamma$$

(11.3)

with branching ratio $1.2 \times 10^{-4}$.

In the above equations, we distinguish between the neutrino associated with muons and the neutrino associated with electrons, for we shall see later in Section 34 that there are two different kinds of neutrinos.

The mass of the neutral pion, $\pi^0$, is 135·0 MeV, which is 4·6 MeV lower than the mass of a charged pion. The $\pi^0$ decays by

$$\pi^0 \rightarrow \gamma + \gamma$$

(11.4)

with branching ratio 98·8 per cent, and by

$$\pi^0 \rightarrow e^+ + e^- + \gamma$$

(11.5)

with branching ratio 1·2 per cent.

To save writing out equations like (11.1) to (11.5), decay modes are summarized by listing the decay products neglecting the charge states. Equations (11.1) are then summarized as $\mu\nu$.

The properties of pions can be investigated by studying reactions in which pions are produced such as

$$p + p \rightarrow p + n + \pi^+$$

(11.6)

and reactions with pions such as

$$\pi^- + p \rightarrow \pi^- + p$$
$$\pi^- + p \rightarrow \pi^0 + n$$
$$\pi^- + p \rightarrow \pi^0 + \pi^- + p$$

(11.7)

A particular combination of bombarding particle and target particle, such as $\pi^- + p$, will in general give rise to various possible final combinations of particles, such as shown in equation (11.7). Each different combination of particles is called a channel.

## References

ALLEN, J. S., *The Neutrino*, 1958. Princeton University Press.

FEYNMAN, R. P., R. B. LEIGHTON and M. SANDS, *Quantum Mechanics*, Vol. III of *The Feynman Lectures on Physics*, 1965. Addison-Wesley, Reading, Mass.

SEGRE, E. and C. E. WIEGAND, 'The antiproton', *Sci. Amer.* June 1956. (Also available as reprint 244, Freeman, San Francisco.)

SNELL, A. H. and F. PLEASONTON, *Phys. Rev.* **97** (1955) 246; **100** (1955) 1396.

### Exercises

1  Show that a photon cannot create an electron–positron pair in free space because of the conservation of energy and momentum.

2  Calculate the threshold for the production of an electron–positron pair by a photon in the presence of an initially stationary electron.

3  Calculate the threshold for production of an electron–positron pair in proton–proton collisions.

4  Draw the lowest order Feynman diagrams for the Compton effect.

5  By considering a time slot of the diagrams obtained in Exercise 4 before and after each vertex, draw diagrams similar to those of Fig. 8.2b, showing the positions and directions of motion of the particles.

6  Draw the Feynman diagram for the inverse $\beta$-decay, equation (9.8).

7  Draw the lowest order Feynman diagrams for
   (a) the scattering of positrons by protons;
   (b) bremsstrahlung production in electron–proton collisions;
   (c) the production of an electron–positron pair by a photon in the presence of an electron;
   (d) electron–positron production in an electron–proton collision.

8  For each of the following decay reactions, give all possible electric charges of each particle and $Q$, the energy released in the decay.
   (a)  $\pi \rightarrow \mu + \nu$
   (b)  $\pi \rightarrow \gamma + \gamma$
   (c)  $\mu \rightarrow e + \nu + \nu$.
   For decay (a), calculate the kinetic energy of the muon in the reference frame in which the initial pion is stationary.

9  In what ways does the $\mu$-meson differ from the meson predicted by Yukawa?

10  Calculate the threshold for antiproton production in proton–proton collisions.

# Properties of the pion

**3**

## 12 The spin of the $\pi^+$

The spin of the $\pi^+$ was determined by applying the principle of detailed balance (Appendix D) to the breakup of a deuteron after absorbing a $\pi^+$,

$$d + \pi^+ \to p + p \tag{12.1}$$

with cross-section $\sigma_{abs}$, and the corresponding inverse reaction,

$$p + p \to d + \pi^+ \tag{12.2}$$

with cross-section $\sigma_{prod}$ (see Appendix C for the definition of cross-section). According to equation (D.16)

$$\tfrac{1}{2}(2J_p + 1)^2 p_p^2 \sigma_{prod} = (2J_d + 1)(2J_\pi + 1)p_\pi^2 \sigma_{abs} \tag{12.3}$$

Using the known spins of the proton and the deuteron, $J_p = \tfrac{1}{2}$ and $J_d = 1$, we have

$$2J_\pi + 1 = \frac{2}{3}\left(\frac{p_p}{p_\pi}\right)^2 \frac{\sigma_{prod}}{\sigma_{abs}}$$

The experimental results show undoubtedly that $J_\pi = 0$ (Cartwright, 1953) (see Exercise 1).

We assume that the $\pi^-$ has the same spin as the $\pi^+$.

## 13 Parity†

Consider the motion of a single particle, such that the physics of this particle is unaltered by inversion of coordinates through the origin

$$\mathbf{r} \to -\mathbf{r} \tag{13.1}$$

A particular example would be a particle moving in a spherically symmetric potential $V(r)$. Consider a quantum mechanical state of the particle described by the wave function $\psi(\mathbf{r})$.

† See Feynman (1965) and Ziock (1969).

If $\psi(\mathbf{r})$ is to be unchanged by the inversion, we must have

$$\psi(-\mathbf{r}) = \psi(\mathbf{r}) \tag{13.2}$$

However, this is too restrictive, because the wave function is not itself physically observable. The probability density for finding the particle at $\mathbf{r}$ is

$$|\psi(\mathbf{r})|^2 = \psi^*(\mathbf{r})\psi(\mathbf{r}) \tag{13.3}$$

and we can only require that this probability density be unaltered by inversion, so that we have

$$\psi(-\mathbf{r}) = e^{i\delta}\psi(\mathbf{r}) \tag{13.4}$$

Performing another inversion through the origin, we have

$$\psi(\mathbf{r}) = e^{i\delta}\psi(-\mathbf{r}) = e^{2i\delta}\psi(\mathbf{r}) \tag{13.5}$$

so that

$$e^{2i\delta} = 1; \qquad e^{i\delta} = \pm 1 \tag{13.6}$$

If

$$\psi(-\mathbf{r}) = +\psi(\mathbf{r})$$

we say the state has even parity or ' + ' parity, and if

$$\psi(-\mathbf{r}) = -\psi(\mathbf{r})$$

we say the state has odd parity or ' − ' parity.

An inversion of coordinates through the origin is represented in quantum mechanics by the operator $P$ where

$$P\psi(\mathbf{r}) = \psi(-\mathbf{r})$$

$P$ is called the parity operator or inversion operator. The eigenvalues of $P$ are $\pm 1$.

$$P\psi(\mathbf{r}) = \pm\psi(\mathbf{r})$$

Inversion of coordinates through the origin is also called the parity transformation.

Parity can also be discussed by considering reflection in a plane, since reflection in a plane is equivalent to inversion through the origin followed by a rotation through 180°, as shown in Fig. 13.1.

Inversion through the origin changes a right-handed coordinate system to a left-handed coordinate system. The invariance of physical laws under inversion is equivalent to those laws being the same in left-handed and right-handed coordinate systems.

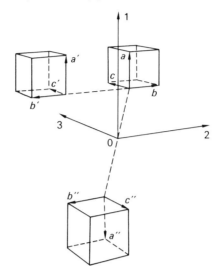

FIGURE 13.1    $a'b'c'$ is the mirror image of $abc$ obtained by reflection in the (1, 3) plane. $a''b''c''$ is the inversion of $abc$ through the origin $O$. $a'b'c'$ can be obtained from $a''b''c''$ by a rotation through 180° about the 2-axis.

For a system which is invariant under inversion, the parity is unchanged in time, or conserved. If such a system initially has even parity, it will have even parity at all times, even though the number of particles in the system may change. By considering a system of two non-interacting particles, 1 and 2, described by a product wave function

$$\psi(\mathbf{r}_1)\phi(\mathbf{r}_2)$$

we see that the parity of the system is the product of the parities of each particle. Parity is a multiplicative quantum number.

Until 1956 it was thought that all the laws of physics were unchanged by a reflection or an inversion of coordinates, and consequently that parity was conserved in all reactions. The possibility of an exception to invariance under reflections was first pointed out by Lee and Yang in 1956, in particular for $\beta$ decay, as will be discussed later in Chapter 7. However, for the time being, we will be restricting our interest to interactions which are invariant under reflections, and consequently conserve parity.

We know that electromagnetic interactions are invariant under reflections and conserve parity. Experiments in nuclear physics show that nuclear forces are invariant under reflections and conserve parity. Since nuclear forces are partly due to the exchange of pions, the interactions of pions with nucleons conserve parity.

As well as the parity due to its spatial state, a particle may have an intrinsic parity. The total parity is then the product of the intrinsic and spatial parity. In processes in which no particles are created or destroyed, the intrinsic parities of particles are unimportant and have no observable consequences. In reactions in which particles are created or destroyed, the intrinsic parities of the particles must be included in determining selection rules due to conservation of parity. The intrinsic parity of a particle only has absolute meaning if the particle can be arbitrarily created or destroyed. Particles that cannot be arbitrarily created or destroyed are assigned intrinsic parities arbitrarily, and the intrinsic parities of other particles into which they can change in reactions are determined by using conservation of parity. The consistency of our assignments of intrinsic parity will be a verification that parity is conserved in the reactions we are considering. This procedure may seem very arbitrary and unworkable. We will meet several examples of this procedure later, and we will see that it does work. Basically, the reason it works is that there are many reactions which conserve parity, and they are easily separated from those that do not. If hardly any reactions conserved parity, the concept of intrinsic parity would be meaningless.

A particle with spin 0 may be described by a scalar wave function, and then has even intrinsic parity. Alternatively, it may have odd intrinsic parity and be described by a pseudoscalar wave function.

A pseudoscalar quantity transforms as a scalar under rotations, but changes sign under reflections. An elementary example of a pseudoscalar quantity is the scalar triple product of three vectors,

$$(\mathbf{A} \times \mathbf{B}) . \mathbf{C}$$

as the sign of this quantity depends on whether a right-handed or left-handed coordinate system is used.

For a wave function $\psi_l(\mathbf{r})$ describing a particle with orbital angular momentum $\hbar l$ moving in a spherically symmetric potential (Saxon, 1968)

$$\psi_l(-\mathbf{r}) = (-1)^l \psi_l(\mathbf{r}) \qquad (13.7)$$

If the particle has even intrinsic parity, then the total parity is

$(-1)^l$. If the particle has odd intrinsic parity, the total parity is $(-1)^{l+1}$.

For two particles interacting through a central potential

$$\psi(\mathbf{r}_1, \mathbf{r}_2) = e^{i\mathbf{K} \cdot \mathbf{R}} \phi_l(\mathbf{r})$$

where $\mathbf{R}$ is the position coordinate of the centre-of-mass, $\hbar\mathbf{K}$ is the total momentum and

$$\mathbf{r} = \mathbf{r}_1 - \mathbf{r}_2$$

$\phi_l(\mathbf{r})$ has the same form as the wave function of a single particle with orbital angular momentum $\hbar l$ (Eisberg, 1961). $\hbar l$ is the orbital angular momentum of the two-particle system in the centre-of-mass frame, and is sometimes called the relative angular momentum. In the centre-of-mass frame, $\mathbf{K} = 0$, and the parity is $(-1)^l$. It is worth noting that, in the centre-of-mass frame, an inversion of coordinates produces the same effect as the interchange of the spatial coordinates of the two particles.

It is not possible to measure the intrinsic parity of a fermion directly as fermions are always created and destroyed in pairs. It is possible to measure the intrinsic parity of a boson, as bosons can be arbitrarily created or destroyed.

## 14 The parity of the $\pi^-$

Consider the capture of $\pi^-$ in deuterium,

$$\pi^- + d \rightarrow n + n \tag{14.1}$$

The $\pi^-$ are first slowed down by losing energy by ionization until they are practically at rest, and then go into atomic orbits around the deuteron. The capture of $\pi^-$ takes place mainly from the $1s$ state, since the $1s$ state has the largest probability density $|\psi|^2$ inside the deuteron.

We arbitrarily assign the same intrinsic parity, $+1$, to the proton and the neutron. The internal orbital angular momentum of the deuteron is even, and so the parity of the deuteron is $+1$. Since the $\pi^-$ is in a $1s$ state, the parity of the system $(\pi^- + d)$ is just the parity of the $\pi^-$, which by conservation of parity will be the parity of the final state consisting of two neutrons.

Since the pion has zero spin and is captured from a $1s$ state, $l = 0$, the total spin of the initial state is the spin of the deuteron which is $J = 1$. By conservation of angular momentum, the final state will have $J=1$. We consider all the possible states of two neutrons for which $J=1$.

For two particles with spins $s_1$ and $s_2$, the total angular momentum $J$ can be obtained by first adding the spins to yield

$$\mathbf{S} = \mathbf{s}_1 + \mathbf{s}_2$$

and then adding $S$ to the orbital angular momentum $l$ to yield $J$

$$\mathbf{J} = \mathbf{S} + \mathbf{l}$$

We have

$$l = 0, 1, 2, \ldots$$

and for

$$s_1 = \tfrac{1}{2}, \qquad s_2 = \tfrac{1}{2}$$
$$S = 0 \text{ or } 1$$

However only the combinations shown in Table 14.1 will give $J=1$. We now consider the requirement that the total wave function must be antisymmetric with respect to interchange of the two neutrons. For $S=0$, the spin function is antisymmetric with respect to the interchange, and for $S=1$ the spin function is symmetric with respect to the interchange, and the spatial wave function has the sign $(-1)^l$ on the interchange of the two neutrons. The only state in Table 14.1 that is totally antisymmetric with respect to interchange of the two neutrons is that with $S=1$ and $l=1$. Since the parity of the system of two neutrons is $(-1)^l = -1$, it follows that the intrinsic parity of the $\pi^-$ is $-1$ (or odd), and the pion is represented by a pseudoscalar wave function.

TABLE 14.1   Possible values of $S$ and $l$ for system of two particles with $J=1$

| $S$ | $l$ |
|---|---|
| 0 | 1 |
| 1 | 0 |
| 1 | 1 |
| 1 | 2 |

## 15 The spin and parity of the $\pi^0$

The determination of the spin and parity of the $\pi^0$ requires the use of sophisticated quantum mechanics beyond the scope of this book, so that only a brief account will be given. A detailed account of the required theory is given by Williams (1961).

From the observed decay

$$\pi^0 \to 2\gamma$$

it can be deduced that the spin $J$ is not 1, and thus it is almost certain that $J=0$. The parity could be determined by measuring the polarization of the photons. For a scalar $\pi^0$, the two photons would have the same plane polarization. For a pseudoscalar $\pi^0$, the two photons have perpendicular polarization planes. Such an experiment has not been performed directly, because of the difficulty of measuring the plane polarization of high energy $\gamma$ rays. However, the study of the angular correlation of the decay products in the decay of the $\pi^0$ into two electron–positron pairs

$$\pi^0 \to e^+ + e^- + e^+ + e^-$$

yields the result that the $\pi^0$ has odd parity (Plano, 1959). We assume that the $\pi^+$ also has odd parity.

## 16  Parity and absolute conservation laws

When discussing the parity of the $\pi^-$, we assumed that the proton and the neutron have the same intrinsic parity. If we assume that the proton and the neutron have opposite intrinsic parities, then we find that the $\pi^-$ and $\pi^0$ have opposite intrinsic parities. It is impossible to determine the intrinsic parity of charged pions independently of that of the nucleon. However, as we have seen, the parity of the $\pi^0$ can be determined unambiguously. Parities are compared by studying reactions. But all reactions conserve charge, and so we cannot compare parities of different charge states. We could alter the parities of all states of charge $+e$ by $-1$ without changing any physics. Then the neutron and proton have opposite intrinsic parities. This only complicates the formalism unnecessarily. So, we choose the parities for different charge states to give the simplest scheme; namely, using $P$ for the parity,

$$P_n = P_p; \qquad P_{\pi^+} = P_{\pi^0} = P_{\pi^-}$$

A similar situation occurs wherever there is a quantity which is absolutely conserved, and the intrinsic parity of some state has to be arbitrarily assigned.

For instance, the fact that the number of nucleons involved in a reaction is always conserved is described by the conservation of baryon number $B$.

The neutron and proton are assigned baryon number $B=1$, and the antineutron and antiproton $B=-1$. The $\pi^0$, $\pi^+$ and $\pi^-$ are assigned $B=0$ (as are also electrons, neutrinos, $\mu$-mesons and photons). Then in all reactions, the baryon number is conserved. For example consider the following reaction with the appropriate baryon numbers shown under each particle symbol

$$p + \bar{p} \rightarrow \pi^+ + \pi^- + \pi^0$$
$$(1) + (-1) = (0) + (0) + (0)$$

We find that the parities of states with different baryon numbers cannot be absolutely compared. The parity of the $\pi^0$ with $B=0$ is uniquely determined as $P_{\pi^0} = -1$. But the parity of the nucleon could be taken as odd, i.e.

$$P_p = P_n = -1$$

and physics would be unaltered. However, since the nucleon is the starting point in the determination of intrinsic parities it is more convenient to take

$$P_n = P_p = +1$$

### References

CARTWRIGHT, W. F., C. RICHMAN, M. N. WHITEHEAD and H. A. WILCOX, *Phys. Rev.* **91** (1953) 677.

EISBERG, R. M., *Fundamentals of Modern Physics*, 1961, Wiley, New York. Chapter 10.

FEYNMAN, R. P., R. B. LEIGHTON and M. SANDS, *Quantum Mechanics*, Vol. III of *The Feynman Lectures on Physics*, 1965. Addison–Wesley, Reading, Mass. Section 17.2.

PLANO, R., A. PRODELL, N. SAMIOS, M. SCHWARTZ and J. STEINBERGER, *Phys. Rev. Letters* **3** (1959) 525.

SAXON, D. S., *Elementary Quantum Mechanics*, 1968. Holden–Day, San Francisco. Chapter IX.

WILLIAMS, W. S. C., *An Introduction to Elementary Particles*, 1961. Academic Press, New York. Section 7.5.

ZIOCK, K., *Basic Quantum Mechanics*, 1969. Wiley, New York. Section 6.7.

## Exercises

1   The cross-section for reaction (12.2) is $\sigma_{\text{prod}} = (0.18 \pm 0.06) \times 10^{-27}$ cm$^2$ for protons with kinetic energy of 340 MeV in the laboratory frame. The cross-section for reaction (12.1) is $\sigma_{\text{abs}} = (3.1 \pm 0.3) \times 10^{-27}$ cm$^2$ when the incident pions have a kinetic energy of 29 MeV in the laboratory frame. The cross-sections vary sufficiently slowly with energy, and the energies of the pions in the centre-of-mass frame are sufficiently close to these two reactions to be compared.

Find the spin of the $\pi^+$ (see Cartwright, 1953). (*Note*: since the answer must be an integer, great accuracy is not needed in the calculation, and non-relativistic kinematics may be used.)

2   In section 14, the determination of the intrinsic parity of the $\pi^-$ was discussed assuming that the intrinsic parity of both neutron and proton was even. Using a similar procedure, what intrinsic parity should be assigned to the $\pi^-$ if the following assignments of intrinsic parity are made, (a) $P_n = +1$, $P_p = -1$; (b) $P_n = -1$, $P_p = +1$; (c) $P_n = P_p = -1$.

3   Calculate the thresholds for the following reactions in the laboratory frame, always assuming the proton is the target:
(a) $p + p \rightarrow p + p + \pi^0$
(b) $p + p \rightarrow p + n + \pi^+$
(c) $p + p \rightarrow p + p + \pi^+ + \pi^-$
(d) $\pi^- + p \rightarrow p + \bar{p} + n$
(e) $\pi^0 + p \rightarrow \pi^+ + n$
(f) $p + p \rightarrow p + p + \pi^+ + \pi^- + \pi^0$
(g) $n + p \rightarrow p + p + \pi^- + \pi^- + \pi^+$.

4   Which of the following processes are absolutely forbidden, and why?
(a) $p \rightarrow e^+ + \gamma$
(b) $n \rightarrow p + e^- + \bar{\nu}_e$
(c) $\pi^0 + \pi^- \rightarrow \bar{n} + p$
(d) $n \rightarrow p + e^+ + \nu_e$
(e) $\pi^0 + n \rightarrow \pi^- + \bar{p}$
(f) $\bar{n} + n \rightarrow \pi^0 + \pi^+ + \pi^-$
(g) $\pi^+ + n \rightarrow \pi^- + p$.

5   Compare the properties of pions and photons.

# Nucleons and pions

# 4

## 17 Isospin

The elementary particles tend to occur in groups of roughly the same mass but with different charges. For instance, the mass of the neutron is roughly that of the proton, and also the mass of a neutral pion is roughly that of a charged pion. When this was known only for the neutron and proton, Heisenberg in 1932 suggested that the proton and the neutron could be regarded as two charge states of a particle which he named the nucleon.

In the theory of atomic spectra, a level which has multiplicity $(2S+1)$, as shown by the Zeeman effect of splitting into $(2S+1)$ levels in a magnetic field, has spin $S$. This spin $S$ can be identified with the angular momentum of the system, and operators for the components of this angular momentum, $S_x, S_y, S_z$, can be defined, and have certain commutation relations

$$[S_x, S_y] = iS_z, \quad [S_y, S_z] = iS_x, \quad [S_z, S_x] = iS_y \quad (17.1)$$

(Strictly the angular momentum is $\hbar S$; it is frequently convenient to omit the $\hbar$, which can be accomplished by choosing units so that $\hbar = 1$.)

The nucleon has multiplicity $2 = (2 \times \frac{1}{2} + 1)$, and by analogy with the theory of atomic spectra, is assigned a quantity called isospin $I = \frac{1}{2}$, to yield multiplicity $2I + 1 = 2$. However, isospin cannot in any way be identified with an angular momentum, and has no relation to the spatial properties of the nucleon. Isospin has nothing to do with ordinary space, but the isospin can be considered to have three components $I_1, I_2, I_3$ along three orthogonal axes in some three-dimensional abstract space, which we call isospin space or charge space. The components of isospin have the same commutation relations as ordinary spin,

$$\begin{aligned} [I_1, I_2] &= iI_3 \\ [I_2, I_3] &= iI_1 \\ [I_3, I_1] &= iI_2 \end{aligned} \quad (17.2)$$

There is no need to consider the abstract charge space at all, as all the properties of isospin can be determined algebraically from the above commutation relations. However, physicists do not like doing mathematics that can be avoided, and by introducing an abstract charge space we can deal with isospin in exactly the same way as we are accustomed to dealing with angular momentum. For instance, as the square of the spin, $\mathbf{S}^2$, has eigenvalues $S(S+1)$, so does the square of the isospin, $\mathbf{I}^2$, have eigenvalues $I(I+1)$. The addition of the isospins of several particles can be dealt with by the vector model, used in the theory of atomic spectra (e.g. Eisberg, 1961).

As an example, consider first the case for ordinary spin, when two spins each of $\frac{1}{2}$ are added. The total spin can be 0 or 1. Similarly, the isospins of two nucleons (each having isospin of $\frac{1}{2}$) can be added to give a total isospin of 0 or 1.

The $(2S+1)$ states of a system with spin $S$ are designated by the $(2S+1)$ different values of the $z$ component $S_z$

$$S_z = -S, -S+1, \cdots, S-1, S$$

Similarly, the $(2I+1)$ states of a system with isospin $I$ are designated by the $(2I+1)$ different values of the component $I_3$

$$I_3 = -I, -I+1, \cdots, I-1, I$$

The direction of the third axis in charge space is chosen so that $I_3 = +\frac{1}{2}$ for the proton and $I_3 = -\frac{1}{2}$ for the neutron.

The pion has three charge states, and so has $I=1$.

Isospin was originally called isotopic spin; it is also called isobaric spin, $i$-spin and $I$-spin.

We now consider in detail the system of two nucleons. The wave function has the form

$$\Psi = \psi(\mathbf{r}_1, \mathbf{r}_2)\, \psi_{\text{spin}}\, \psi_{\text{isospin}}$$

where $\psi_{\text{spin}}$ and $\psi_{\text{isospin}}$ depend only on the spin and isospin respectively. We consider the case where $\psi_{\text{spin}}$ is an eigenfunction of $\mathbf{S}^2$ and $S_z$, and $\psi_{\text{isospin}}$ is an eigenfunction of $\mathbf{I}^2$ and $I_3$, where $\mathbf{S}$ and $\mathbf{I}$ are respectively the total spin and the total isospin of the system,

$$\mathbf{S} = \mathbf{s}(1) + \mathbf{s}(2) \tag{17.3}$$

$$\mathbf{I} = \mathbf{i}(1) + \mathbf{i}(2) \tag{17.4}$$

where $\mathbf{s}(k)$ and $\mathbf{i}(k)$ are the spin and isospin of the $k$th nucleon.

We revise briefly the treatment of spin before applying the same treatment to isospin. For the $k$th nucleon, we introduce $\chi_+^s(k)$, $\chi_-^s(k)$

which are eigenfunctions of $s^2(k)$, $s_z(k)$,

$$s_z(k) \, \chi_+^s(k) = +\tfrac{1}{2}\chi_+^s(k)$$
$$s_z(k) \, \chi_-^s(k) = -\tfrac{1}{2}\chi_-^s(k) \tag{17.5}$$

The spin states of two nucleons which are symmetric in the spins are the three states with $S=1$,

$$
\begin{aligned}
S_z &= +1 & & \chi_+^s(1)\chi_+^s(2) \\
S_z &= 0 & & 2^{-\frac{1}{2}}\{\chi_+^s(1)\chi_-^s(2)+\chi_-^s(1)\chi_+^s(2)\} \\
S_z &= -1 & & \chi_-^s(1)\chi_-^s(2)
\end{aligned}
$$

and the spin state antisymmetric in the spins is the state with $S=0$,

$$2^{-\frac{1}{2}}\{\chi_+^s(1)\chi_-^s(2)-\chi_-^s(1)\chi_+^s(2)\}$$

Similarly we introduce $\chi_+^i(k)$, $\chi_-^i(k)$ which are eigenstates of $i^2(k)$, $i_3(k)$

$$i_3(k)\chi_+^i(k) = +\tfrac{1}{2}\chi_+^i(k)$$
$$i_3(k)\chi_-^i(k) = -\tfrac{1}{2}\chi_-^i(k) \tag{17.6}$$

$\chi_+^i$ represents a proton, and $\chi_-^i$ represents a neutron. The isospin states of two nucleons are given in Table 17.1. We see that there is one state of two protons, one state of two neutrons, and two states of neutron and proton.

TABLE 17.1  Isospin states of two nucleons

| Type of symmetry | $I$ | $I_3$ | $\psi_{\text{isospin}}$ | Content |
|---|---|---|---|---|
| Symmetric | 1 | +1 | $\chi_+^i(1)\chi_+^i(2)$ | $pp$ |
| | | 0 | $2^{-\frac{1}{2}}\{\chi_+^i(1)\chi_-^i(2)+\chi_-^i(1)\chi_+^i(2)\}$ | $np$ |
| | | −1 | $\chi_-^i(1)\chi_-^i(2)$ | $nn$ |
| Antisymmetric | 0 | 0 | $2^{-\frac{1}{2}}\{\chi_+^i(1)\chi_-^i(2)-\chi_-^i(1)\chi_+^i(2)\}$ | $np$ |

For simplicity we work in the centre-of-mass system and consider a state with orbital angular momentum $L=0$. Then

$$\Psi = \psi(r)\psi_{\text{spin}}\psi_{\text{isospin}} \tag{17.7}$$

where $r$ is the distance between the two nucleons. $\psi(r)$ is symmetric with respect to interchange of two nucleons. Treating the two nucleons as identical fermions, we must select products $\psi_{\text{spin}}\psi_{\text{isospin}}$ that are antisymmetric in interchange of the two nucleons in order that the total wave function will be antisymmetric. There are six possible antisymmetric states as shown in Table 17.2.

TABLE 17.2    Spin–isospin states of two nucleons with $L=0$

| Spin state | Isospin state | Number of states |
|---|---|---|
| Symmetric<br>$S = 1$<br>3 states, $S_z = 0, \pm 1$ | Antisymmetric<br>$I = 0$<br>1 state, $I_3 = 0$ | 3 |
| Antisymmetric<br>$S = 0$<br>1 state, $S_z = 0$ | Symmetric<br>$I = 1$<br>3 states, $I_3 = 0, \pm 1$ | 3 |
|  |  | Total   6 |

We wish to show there are the same number of possible states when the same system of two nucleons with $L=0$ is treated regarding the neutron as distinguishable from the proton. The total wave function is now

$$\Psi = \psi(r)\psi_{\text{spin}} \qquad (17.8)$$

and we must distinguish the three cases where the system is *nn*, *pp* or *np*.

(a) *nn* case. The system consists of two identical fermions, and so $S=0$ to ensure antisymmetry on interchange of the two neutrons. 1 state.

(b) *pp* case. Similarly $S=0$ only. 1 state.

(c) *np* case. The two particles are distinguishable, so there is no symmetry requirement on the wave function. Both $S=0$ and $S=1$ states are permitted. 4 states.

There are altogether six possible states, in agreement with the number of states in the isospin scheme. We have considered only the $L=0$ case, but it can be shown that the enumeration of states in the isospin scheme is always consistent with treating neutrons and protons separately. (See Exercise 2.)

## 18  Charge independence of nuclear forces

Experiments in nuclear physics show that, to a good approximation, nuclear forces are charge independent.

In the $I=1$ state there are three possible forces to consider corresponding to the three values of $I_3$,

$$I_3 = +1 \qquad p\text{--}p \text{ force}$$
$$I_3 = \phantom{+}0 \qquad n\text{--}p \text{ force}$$
$$I_3 = -1 \qquad n\text{--}n \text{ force.}$$

Experimentally, these three forces have been found to be the same to a good approximation.

In the $I=0$ state, there is only one possible force,

$$I_3 = \phantom{+}0 \qquad n\text{--}p \text{ force}$$

The $n$–$p$ force in the isospin state $I=0$ is found to be different from the $n$–$p$ force in the $I=1$ state.

This can be summed up by saying that the nuclear force is invariant under rotations in isospin space. The nuclear force does not depend on $I_3$, but can, and in fact does, depend on $I$, the magnitude of the isospin, which is a scalar in isospin space.

The deuteron is a bound state of neutron and proton with $S=1$ and $L=0$ and so $I=0$, since quite generally $L+S+I$ must be odd for a two-nucleon system, to satisfy the exclusion principle. (See Exercise 1.) Thus a bound state of two nucleons occurs with $S=1$, $I=0$.

There is no bound state of the deuteron with $S=0$, $I=1$, and this implies that there should be no bound states of two neutrons or of two protons.

Because of charge independence of nuclear forces, the total isospin is a good quantum number not only for two nucleons, but for many nucleons, and so is a useful quantum number in nuclear physics. However, there is not complete invariance under rotations in isospin space, as this invariance is broken by the electromagnetic interaction. Since protons are electrically charged, the interaction between two protons differs from that between two neutrons by the additional Coulomb repulsion. However, neglecting electromagnetic effects, nuclear forces are invariant under rotations in isospin space, isospin is a good quantum number, and the isospin of a system stays constant in time so that isospin is conserved.

## 19 Isospin of pions

We now consider the three pions $\pi^-$, $\pi^0$, $\pi^+$ as a charge multiplet or isospin multiplet of multiplicity $2I+1=3$ yielding $I=1$. We identify the $I_3 = +1$ state as the $\pi^+$; $I_3 = 0$ as the $\pi^0$; $I_3 = -1$ as the $\pi^-$. This identification is not arbitrary, as we shall see below. Since isospin is conserved in the interaction of nucleons, we wish to postulate conservation of isospin in the pion–nucleon interaction, since the nucleon–nucleon interaction is in part due to the exchange of virtual pions. Alternative identifications of the states of different $I_3$ with the various pions are inconsistent with the conservation of $I_3$. Consider the virtual reaction

$$n \rightarrow p + \pi^-$$

$(p + \pi^-)$ has $I_3 = -\frac{1}{2}$ which is the same $I_3$ as the neutron. With this choice ($I_3 = -1$ for $\pi^-$) the conservation of $I_3$ follows from the conservation of charge and the conservation of baryon number.

We can write $Q$, the charge in units of $e$, as

$$Q = I_3 + \frac{B}{2} \tag{19.1}$$

where $B$ is the baryon number. Since the antiparticle of a particle of charge $Q$ and baryon number $B$, has charge $-Q$ and baryon number $-B$, it must also have third component of isospin $-I_3$ where $I_3$ is the isospin of the corresponding particle. (See Exercise 3.)

The total interaction among nucleons and pions is not completely invariant under rotations in isospin space because of electromagnetic effects, and so isospin is only approximately conserved. Note however, that even with electromagnetic effects included, $I_3$ is conserved since its conservation follows from conservation of charge and baryon number.

In reactions proceeding mainly by the electromagnetic interaction, such as emission of a photon or production of an electron–positron pair, isospin is not conserved. Consequently, since the strongest interaction of electrons is electromagnetic, it is meaningless to define an isospin for the electron. Similarly, it is meaningless to define an isospin for a muon.

As an example of a prediction made on the basis of the conservation of isospin, consider the two reactions

$$n + p \rightarrow d + \pi^0 \tag{19.2}$$

$$p + p \rightarrow d + \pi^+ \tag{19.3}$$

The deuteron has $I = 0$. Thus the final state, for both reactions, has $I = 1$. For $(p + p)$, $I = 1$ since $I_3 = +1$. $(n + p)$ can have either $I = 0$ or $I = 1$. From Table 17.1, solving for $\chi^i_-(1)\chi^i_+(2)$ which represents $(n + p)$, we have

$$\chi^i_-(1)\chi^i_+(2) = \frac{1}{\sqrt{2}}\{\psi_{\text{isospin}}(I = 1, I_3 = 0) - \psi_{\text{isospin}}(I = 0, I_3 = 0)\}$$

$$\tag{19.4}$$

i.e. $(n+p)$ is a superposition of equal amounts of $I=1$ and $I=0$ states. Since only the $I=1$ state contributes to the reaction, the $(n+p)$ reaction should have only half the cross-section of the $(p+p)$ reaction at the same energy (neglecting electromagnetic effects which should be small). This prediction is confirmed by the experimental results, to within the rather large experimental error. The cross-section for reaction 2 has been measured by Fliagin *et al.* (1959) and the results and comparison of reactions 2 and 3 are shown in Table 19.1.

TABLE 19.1   From FLIAGIN, V. B., V. P. DZHELEPOV, V. S. KISELEV and K. O. OGANESIAN, *Soviet Physics JETP* **35**(8) (1959) 592

| Energy of incident nucleon (MeV) | Reaction | Angular distribution | Total cross-section, $\sigma$ ($10^{-27}$ cm$^2$) | Experiment |
|---|---|---|---|---|
| 580 | $p+p \to \pi^+ +d$ | $(0.216 \pm 0.033)$ $+\cos^2\theta$ | $3.10 \pm 0.24$ | 1 |
| 600 | $n+p \to \pi^0 +d$ | $(0.220 \pm 0.022)$ $+\cos^2\theta$ | $1.5 \pm 0.3$ | Fliagin (1959) |
| 660 | $p+p \to \pi^+ +d$ | $(0.23 \pm 0.30)$ $+\cos^2\theta$ | $3.1 \pm 0.2$ | Fliagin (1959) |
| 610 | $p+p \to \pi^+ +d$ | | $3.15 \pm 0.22$ | 2 |

Refs.   1. Cohn, C. E., *Phys. Rev.* **105** (1957) 1582.
2. Meshcheriakov, M. G. and B. S. Neganov, *Dokl. Akad. Nauk. SSSR* **100** (1955) 677.

### References

EISBERG, R. M., *Fundamentals of Modern Physics*, 1961. Wiley, New York.

FLIAGIN, V. B., V. P. DZHELEPOV, V. S. KISELEV and K. O. OGANESIAN, *Soviet Physics JETP* **35**(8) (1959) 592. In Russian, *J. Exptl. Theoret. Phys.* (U.S.S.R.) **35** (1958) 854.

### Exercises

1   Show that $L+S+I$ must be odd for a two nucleon system, in order to satisfy the Pauli exclusion principle.

2   Enumerate the states of a system of two nucleons with $L=1$
(a) using the isospin scheme,
(b) treating the neutron as distinguishable from the proton.
(See Section 17 for treatment of the $L=0$ case.)

3   Check the expression

$$Q = I_3 + \tfrac{1}{2}B$$

for $p$, $n$, $\bar{p}$, $\bar{n}$, $\pi^+$, $\pi^-$, $\pi^0$.

4   (a) What are the possible values of isospin for a system of a nucleon and a pion?
(b) What is the isospin of a system of a proton and a $\pi^+$?

5   What are the values of the isospin of all $2\pi$ systems?

# Magnetic moments

5

## 20 Nucleon magnetic moments

To consider a spinning particle classically, we must suppose the particle to have a small but finite extent. Then classically, a spinning electrically charged particle will have a magnetic moment. The spinning charge acts as a small current loop.

In quantum mechanics also, a charged particle with spin has a magnetic moment. According to the Dirac equation, which describes particles with spin $\frac{1}{2}$, the magnetic moment of a particle with spin $\frac{1}{2}$ will be

$$\frac{e\hbar}{2Mc}$$

$(e\hbar/2M_p c)$ is called a nuclear magneton where $M_p$ is the mass of the proton. The magnetic moments of the proton and neutron are found experimentally to be

$$\mu_p = +2\cdot7928 \text{ nuclear magnetons}$$
$$\mu_n = -1\cdot9131 \text{ nuclear magnetons}$$

According to the theory of Dirac, we would expect $\mu_p = 1$ and $\mu_n = 0$.

This discrepancy is thought to be due, at least in part, to the virtual dissociation of nucleons, involving charged mesons. Let us consider the virtual production of a $\pi^+$ by a proton, as shown in the Feynman diagram, Fig. 20.1.

The $\pi^+$ has zero spin and thus no magnetic moment, but it can contribute to the overall magnetic moment of the proton because of its circulation about the neutron. This process must conserve total angular momentum and parity. We will take the orbital angular momentum of the proton to be $l=0$, and take the z-axis along the spin of the proton. Then initially, the total angular momentum is

$$J = \frac{1}{2}, \qquad J_z = +\frac{1}{2}$$

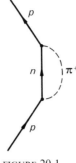

FIGURE 20.1.

and the parity

$$P = P_p = +1$$

The parity of the intermediate state of neutron and pion is $P_n P_\pi (-1)^L = (-1)^{L+1}$ where $L$ is the total angular momentum in the centre-of-mass frame. $P_\pi = -1$, $P_n = +1$. Conservation of parity ensures that $L$ is odd, thus excluding $L = 0$. Since

$$\mathbf{J} = \mathbf{L} + \mathbf{S}$$

and $S$ is the spin of the neutron, namely $\frac{1}{2}$, and $J = \frac{1}{2}$, $L$ must be 1. The intermediate state with $J = \frac{1}{2}$, $J_z = \frac{1}{2}$ consists of contributions from the state $(L_z = 1, S_z = -\frac{1}{2})$ and the state $(L_z = 0, S_z = \frac{1}{2})$. There is no contribution from $L_z = -1$. Thus there is a nett circulation of the $\pi^+$ around the neutron in the same way as the proton was originally spinning. This produces a magnetic moment in the same direction as that of the original proton. The magnetic moment due to the orbital motion of a particle is

$$\frac{e\hbar}{2Mc} L_z$$

Since $M_{\pi^+} \ll M_p$, the effect of the virtual production of a $\pi^+$ is to increase the proton's magnetic moment. And so, we expect $\mu_p > 1$, as observed.

Similarly we may consider the virtual dissociation of a neutron into a proton and a negative pion as shown by Fig. 20.2. As before, we find that there is a nett circulation of the $\pi^-$ in the same way as the neutron was originally spinning. There is a small contribution to the magnetic moment due to the nett spin of the proton, but the bulk of the magnetic moment comes from the circulation of the $\pi^-$,

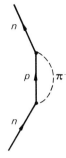

FIGURE 20.2.

because of its smaller mass. This contribution is negative because of the negative charge of the $\pi^-$. And so we expect $\mu_n < 0$, as observed.

In terms of virtual production of $\pi$-mesons, this is about as far as one can go towards explaining nucleon magnetic moments. Even elaborate calculations give no more than $\mu_p > 1$, $\mu_n < 0$. An adequate theory of nucleon magnetic moments is still lacking. We will study other particles later, and virtual dissociation of the nucleon into these other particles also contributes to the nucleon magnetic moments. However, as these other particles are heavier than the pion, we would not expect their contribution to alter the qualitative conclusion, $\mu_p > 1$, $\mu_n < 0$.

## 21 Anomalous magnetic moments of electron and muon

Remembering that the electron takes part in virtual interactions with the electromagnetic field, such as emitting and absorbing virtual photons, we might expect that there would be a small correction to the magnetic moment of the electron. This small correction can be calculated using quantum electrodynamics, and has also been measured.

The magnetic moment $\mu$ associated with an angular momentum $\hbar s$ of a particle of mass $M$ can be written as

$$\mu = g \frac{e\hbar}{2Mc} s \qquad (21.1)$$

where $g$ is called the $g$-factor, and was first used in the theory of atomic spectra. (See for instance Eisberg, 1961.) According to the Dirac equation, the $g$-factor of a spin-$\frac{1}{2}$ particle is

$$g = 2$$

When the effects described by quantum electrodynamics are included, the *g*-factor of the electron differs slightly from 2, and it is convenient to write

$$g = 2(1+a) \tag{21.2}$$

Similarly the *g*-factor of the muon differs from 2 because of virtual interactions of the muon with the electromagnetic field. When considering the effect of the emission of virtual photons by a muon, the production of virtual electron–positron pairs by these virtual photons must also be considered.

$a \equiv \frac{1}{2}(g-2)$ can be measured for electrons and for muons in what is referred to as a $(g-2)$ experiment. When a charged particle with a *g*-factor of exactly 2 moves through a magnetic field which is constant in time, the direction of motion and the direction of the spin of the particle change at the same rate. A beam of such particles which are initially longitudinally polarized, i.e. with spins along the direction of motion, retain their longitudinal polarization. For particles with $g \neq 2$, the type of polarization of an initially longitudinally polarized beam changes with time, and $(g-2)$ can be determined by measuring the change of direction of polarization.

The non-relativistic equations of motion of a particle with spin in a magnetic field are

$$M\frac{d\mathbf{v}}{dt} = \frac{e}{c}\mathbf{v} \times \mathbf{H} \tag{21.3}$$

$$\frac{d(\hbar\mathbf{s})}{dt} = \boldsymbol{\mu} \times \mathbf{H} \tag{21.4}$$

Equation (21.4) states that the rate of change of angular momentum is equal to the torque. Although the spin is essentially quantum mechanical, the classical equation of motion, equation (21.4), can be used to describe the precession of the spin about the direction of the magnetic field, as the expectation value of a quantum mechanical quantity obeys the corresponding classical equation of motion. An explicit quantum mechanical treatment of the precession of spin about the magnetic field is given by Feynman (1965). Substituting (21.1) in (21.4),

$$\frac{d\mathbf{s}}{dt} = \frac{ge}{2Mc}\mathbf{s} \times \mathbf{H} \tag{21.5}$$

We resolve the spin **s** into components along the velocity and perpendicular to the velocity,

$$\mathbf{s} = |\mathbf{s}|(\hat{\mathbf{v}} \cos \phi + \hat{\mathbf{n}} \sin \phi) \qquad (21.6)$$

$\hat{\mathbf{v}}$ and $\hat{\mathbf{n}}$ are unit vectors. $\mathbf{v}.\hat{\mathbf{n}}=0$. From equation (21.3),

$$\dot{\hat{\mathbf{v}}} = \frac{e}{Mc} \hat{\mathbf{v}} \times \mathbf{H} \qquad (21.7)$$

Substituting (21.6) into (21.5), and using (21.7), we obtain

$$\hat{\mathbf{n}}\dot{\phi} \cos \phi + \dot{\hat{\mathbf{n}}} \sin \phi - \hat{\mathbf{v}}\dot{\phi} \sin \phi =$$

$$(g-2) \frac{e}{2Mc} \cos \phi \, \hat{\mathbf{v}} \times \mathbf{H} + \frac{ge}{2Mc} \sin \phi \, \hat{\mathbf{n}} \times \mathbf{H} \qquad (21.8)$$

For a particle with the spin pointing along the direction of motion, $\phi=0$, and

$$\hat{\mathbf{n}}\dot{\phi} = (g-2) \frac{e}{2Mc} \hat{\mathbf{v}} \times \mathbf{H} \qquad (21.9)$$

The particle acquires a transverse polarization perpendicular to the magnetic field.

For arbitrary spin direction, we take the scalar product of both sides of equation (21.8) with $\hat{\mathbf{n}}$, and using $\hat{\mathbf{n}}.\hat{\mathbf{n}}=0$ and $\hat{\mathbf{n}}.\hat{\mathbf{v}}=0$,

$$\dot{\phi} = (g-2) \frac{e}{2Mc} \hat{\mathbf{n}}.(\hat{\mathbf{v}} \times \mathbf{H})$$

$$= (g-2) \frac{e}{2Mc} \hat{\mathbf{v}}.(\mathbf{H} \times \hat{\mathbf{n}}) \qquad (21.10)$$

Equation (21.10) has been obtained relativistically by Bargmann, Michel and Telegdi (1959). For a beam of particles, $\dot{\phi}$ is the rate at which a longitudinal polarization changes into a transverse polarization.

A particularly simple geometry is for a beam of particles moving in a direction perpendicular to the magnetic field and with spin perpendicular to the magnetic field. Then

$$\dot{\phi} = (g-2) \frac{e}{2Mc} \qquad (21.11)$$

$(g-2)$ experiments for the electron and muon have been reviewed recently by Farley (1969). The results of experiment and theory have been summarized by Pipkin (1970) as follows:

$$a_{\text{electron}}(\text{experiment}) = (115\,964\,4 \pm 7) \times 10^{-9}$$
$$a_{\text{electron}}(\text{theory}) \quad = (115\,964\,3 \cdot 6 \pm 2 \cdot 3) \times 10^{-9}$$
$$a_{\text{muon}}(\text{experiment}) \quad = (116\,616 \pm 31) \times 10^{-8}$$
$$a_{\text{muon}}(\text{theory}) \quad = (116\,587 \cdot 2 \pm 2 \cdot 2) \times 10^{-8}$$

An account of earlier experiments measuring $(g-2)$ for the muon is given by Penman (1961).

In contrast to the nucleon magnetic moments, for which we have no exact theory, quantum electrodynamics provides a very accurate description of the magnetic moments of the electron and the muon.

### References

BARGMANN, V., L. MICHEL and V. L. TELEGDI, *Phys. Rev. Letters* **2** (1959) 435.

EISBERG, R. M., *Fundamentals of Modern Physics*, 1961. Wiley, New York. Chapter 13.

FARLEY, F. J. M. *Rivista de Nuovo Cimento* **1** (1969) special number. 59.

FEYNMAN, R. P., R. B. LEIGHTON and M. SANDS, *Quantum Mechanics*, Vol. III of *The Feynman Lectures on Physics*, 1965. Addison-Wesley, Reading, Mass. Section 7.5.

PENMAN, S., 'The muon', *Sc. Amer.*, July 1961. (Also available as reprint 275, Freeman, San Francisco.)

PIPKIN, F. M., *Essays in Physics* **2** (1970)1.

### Exercise

1   What is the total change in $\phi$ for a $\mu$-meson after traversing a circular orbit 1000 times in a uniform magnetic field?

# Strange particles

# 6

## 22 Summary of known particles until 1947

Table 22.1 summarizes the properties of the particles known to exist in 1947, and shows a relatively simple state of affairs compared to the multitude of particles discovered since then. The photon, electron, proton and neutron are all familiar from atomic and nuclear physics. The pions were needed to account for nuclear forces. The neutrino fitted well into the theory of $\beta$ decay. The only particle with no apparent reason for its existence was the muon. The significance of the muon was, and still is, a mystery.

TABLE 22.1   Known particles up to 1947

*Fermions*

| | Particles | | | | | | Antiparticles | | | | |
|---|---|---|---|---|---|---|---|---|---|---|---|
| | $J$ | $I$ | $I_3$ | $Q$ | | | $J$ | $I$ | $I_3$ | $Q$ | |
| $n$ | $\frac{1}{2}$ | $\frac{1}{2}$ | $-\frac{1}{2}$ | $0$ | unstable | $\bar{n}$ | $\frac{1}{2}$ | $\frac{1}{2}$ | $+\frac{1}{2}$ | $0$ | unstable |
| $p$ | $\frac{1}{2}$ | $\frac{1}{2}$ | $+\frac{1}{2}$ | $+1$ | stable | $\bar{p}$ | $\frac{1}{2}$ | $\frac{1}{2}$ | $-\frac{1}{2}$ | $-1$ | stable |
| $\mu$ | $\frac{1}{2}$ | not defined | | $-1$ | unstable | $\bar{\mu}$ | $\frac{1}{2}$ | not defined | | $+1$ | unstable |
| $e$ | $\frac{1}{2}$ | not defined | | $-1$ | stable | $\bar{e}$ | $\frac{1}{2}$ | not defined | | $+1$ | stable |
| $v$ | $\frac{1}{2}$ | not defined | | $0$ | stable | $\bar{v}$ | $\frac{1}{2}$ | not defined | | $0$ | stable |

*Bosons*

| | $J$ | $I$ | $I_3$ | $Q$ | |
|---|---|---|---|---|---|
| $\pi^+$ | $0$ | $1$ | $+1$ | $+1$ | unstable |
| $\pi^0$ | $0$ | $1$ | $0$ | $0$ | unstable |
| $\pi^-$ | $0$ | $1$ | $-1$ | $-1$ | unstable |
| $\gamma$ | $1$ | not defined | | $0$ | stable |

## 23 Strange particles

In 1947, G. D. Rochester and C. C. Butler obtained two cloud chamber photographs of previously unidentified particles in showers of penetrating cosmic-ray particles. These two photographs are shown in Figs. 23.1 and 23.2. The forked track, *ab*, in Fig. 23.1 was due to the decay of a heavy neutral particle, of mass around 1000 electron masses, into two charged particles.

In Fig. 23.2 the bent track was due to the decay of a charged particle of mass around 1000 electron masses into a neutral and a secondary charged particle.

These particles at first were called V-particles because of the appearance of the tracks by which they were detected, and later were called strange particles. Considerable investigation of V-particles using cosmic rays followed, joined by the investigation of V-particles produced in the laboratory from 1953, when the cosmotron, a high-

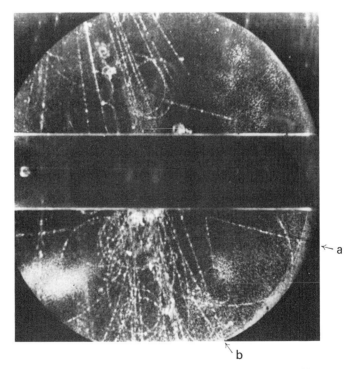

← a

↖ b

FIGURE 23.1   Photograph of cloud chamber tracks obtained by Rochester and Butler (1947) showing decay of a heavy neutral particle. First example of decay $K^0 \rightarrow \pi^+ + \pi^-$.

energy accelerator at the Brookhaven National Laboratory, came into operation, to be later followed by other accelerators.

The strange particles fell into two main groups. One consisted of particles heavier than the nucleon which yielded nucleons when they decayed, and were called hyperons. The symbols $\Lambda$, $\Sigma$, $\Xi$ are used for the various hyperons. Since a hyperon decays eventually to a proton, a hyperon is a baryon and has baryon number one. The hyperons also have antiparticles with baryon number $-1$. The hyperons have spin $\frac{1}{2}$, and are fermions. The other group of strange particles are bosons with spin 0 and are called $K$-mesons or kaons.

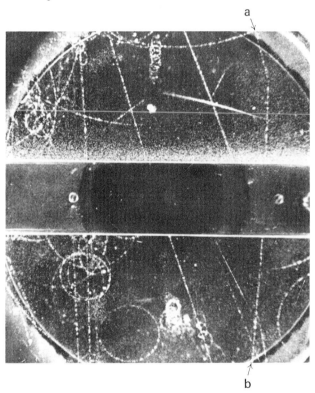

FIGURE 23.2　Photograph of cloud chamber tracks obtained by Rochester and Butler (1947) showing decay of a heavy charged particle. First example of decay $K^+ \rightarrow \mu^+ + \nu$.

## 24 Associated production and strangeness

A typical strange particle is the $\Lambda^0$, an uncharged particle which decays with a mean lifetime of $2 \cdot 5 \times 10^{-10}$ s, the main mode of decay being

$$\Lambda^0 \to p + \pi^- \qquad (24.1)$$

The 'interaction time' for reactions involving nucleons and pions is given approximately by the time for a pion, with a velocity of about the velocity of light, to travel a distance equal to the range of nuclear forces. This 'interaction time' is approximately $10^{-23}$ s, very much shorter than the decay lifetime of the $\Lambda^0$ or of the other strange particles. On the other hand, the rate at which $\Lambda^0$'s and other strange particles are produced is consistent with an 'interaction time' of about $10^{-23}$ s.

In order to explain the fact that strange particles are so readily produced and yet decay so slowly, Pais in 1952 put forward the suggestion of associated production of strange particles. In strong interactions, which are interactions with a strength comparable to the interaction between nucleons or between pions and nucleons, the strange particles were assumed to be produced in groups of two. The reactions in which only one strange particle participates, such as decay, proceed by weak interactions similar to $\beta$ decay or the decay of muons or charged pions. This idea of associated production of strange particles was confirmed by experiment, as for example in the cloud chamber photograph shown in Fig. 24.1a of the reaction

$$\pi^- + p \to K^0 + \Lambda^0 \qquad (24.2)$$

which proceeds via a strong interaction followed by the decays

$$K^0 \to \pi^+ + \pi^- \qquad (24.3)$$

$$\Lambda^0 \to \pi^- + p \qquad (24.4)$$

proceeding via weak interactions. The electrically neutral $K^0$ and $\Lambda^0$ do not make tracks in the cloud chamber, but their existence was inferred from the relationships between the reaction points B, C and D, and from the conservation of energy and momentum.

In 1953, Gell-Mann and Nishijima showed that the associated production of strange particles could be explained by introducing a new additive quantum number called strangeness, and postulating that strangeness was conserved in strong interactions. Then, for instance, two strange particles of opposite strangeness could be produced through the strong interaction in a collision of a pion with a nucleon. Since strangeness is not conserved in the subsequent decay of each strange particle, the decays are attributed to the weak interactions and are therefore slow.

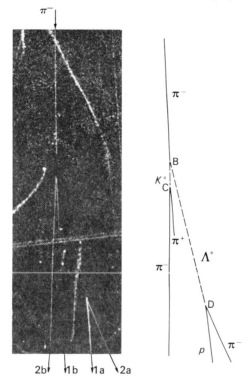

FIGURE 24.1   A 1·5 BeV $\pi^-$ producing a $K^0$ and a $\Lambda^0$ in a collision with a proton. The cloud chamber photograph (a) is from Fowler *et al.* (1954).

The strangeness $S$ of a particle which can be involved in strong interactions is defined by the relation

$$Q = I_3 + \frac{B}{2} + \frac{S}{2}$$  (24.5)

For example, for $S=0$, equation (24.5) reduces to

$$Q = I_3 + \frac{B}{2}$$  (24.6)

which is the relationship we already have for nucleons and $\pi$-mesons. Therefore nucleons and pions have strangeness $S=0$.

Consider the $\Lambda^0$ particle already discussed. The $\Lambda^0$ has a mass of 1116 MeV, and there are no charged particles with masses near this.

So that the multiplicity of the $\Lambda$ is

$$1 = 2I + 1$$

giving $I = 0$; the $\Lambda$ is an isospin singlet. From the decay, equation (24.4), since baryon number is conserved in all reactions, the $\Lambda^0$ has baryon number $B = 1$. $B = 1$, $I_3 = 0$, $Q = 0$ in equation (24.5) yields $S = -1$ for the strangeness of the $\Lambda^0$. Looking at equation (24.4) for the decay of the $\Lambda^0$, we see that the L.H.S. has $S = -1$ whereas the R.H.S. has $S = 0$; so that the decay of the $\Lambda^0$ does not conserve strangeness.

Since the reaction given by equation (24.2) is strong and so conserves strangeness, we can deduce the strangeness of the $K^0$-meson by conservation of strangeness to be

$$S = 1$$

Since the $K^0$-meson has baryon number $B = 0$, from equation (24.5) we find the third component of isospin of the $K^0$ to be

$$I_3 = -\tfrac{1}{2}$$

## 25  *K*-mesons

Since $K$-mesons occur with charges $+e$, $-e$ and $0$ it might seem that the $K$-mesons could form an isospin triplet with $I = 1$. But this is inconsistent with the above result of the $K^0$ having $I_3 = -\frac{1}{2}$, which requires that $I$ is half an odd integer.

A consistent scheme is obtained by regarding $K^+$ and $K^0$ as an isospin doublet with $I = \frac{1}{2}$. $K^+$ has $I_3 = +\frac{1}{2}$, and $K^0$ has $I_3 = -\frac{1}{2}$. The $K^0$ and $K^+$ each have $S = +1$.

The $K^-$ is the antiparticle of the $K^+$, and has $S = -1$, $I = \frac{1}{2}$, $I_3 = -\frac{1}{2}$.

Then the $K^0$ must have an antiparticle $\overline{K^0}$ with $S = -1$, $I = \frac{1}{2}$, $I_3 = +\frac{1}{2}$. There must be two distinct kinds of neutral kaons. This is borne out by there being two different decay constants for the neutral kaon; the decay of neutral $K$-mesons is not a pure exponential decay (as will be discussed more fully in Chapter 9).

The properties of $K$-mesons are summarized in Table 25.1.

The difference between the isospin assignments for $\pi$-mesons and $K$-mesons should be borne in mind. The $\pi$-mesons form an isospin triplet, with

$$\overline{\pi^+} = \pi^-$$
$$\overline{\pi^0} = \pi^0$$

TABLE 25.1   Quantum numbers of $K$-mesons

|  |  | $K^+$ | $K^0$ | $\overline{K^0}$ | $K^-$ |
|---|---|---|---|---|---|
| Baryon number | $B$ | 0 | 0 | 0 | 0 |
| Isospin | $I$ | $\frac{1}{2}$ | $\frac{1}{2}$ | $\frac{1}{2}$ | $\frac{1}{2}$ |
| Third component of isospin | $I_3$ | $+\frac{1}{2}$ | $-\frac{1}{2}$ | $+\frac{1}{2}$ | $-\frac{1}{2}$ |
| Strangeness | $S$ | $+1$ | $+1$ | $-1$ | $-1$ |

i.e. the $\pi^-$ is the antiparticle of the $\pi^+$, and the $\pi^0$ is its own anti-particle. The $K$-mesons form two isospin doublets with

$$\overline{K^+} = K^-$$
$$\overline{K^0} \neq K^0$$

The antiparticle of the $K^0$ is a distinct particle from the $K^0$.

At first sight, it may seem confusing for neutral bosons that there are bosons which have distinct antiparticles as well as bosons which are their own antiparticles. However, it is easily seen that both types of neutral bosons must occur by considering neutral bosons composed of pairs of fermions. A boson composed of a fermion and its corresponding antifermion must be its own antiparticle. An example is positronium which is a bound state of a positron and an electron. A boson composed of two fermions, such that one fermion is not the antiparticle of the other, must have an antiparticle composed of the corresponding two antifermions. An example of such a neutral boson is the hydrogen atom, consisting of a bound state of a proton and an electron.

## 26 Hyperons

The hyperons consist of the $\Lambda$, $\Sigma$ and $\Xi$ particles, all having spin $J = \frac{1}{2}$ and baryon number $B = 1$. The $\Sigma$ has strangeness $S = -1$ and is an isospin triplet, $I = 1$, with particles $\Sigma^-$, $\Sigma^0$ and $\Sigma^+$. The $\Xi$ has $S = -2$ and occurs as $\Xi^0$ and $\Xi^-$ forming an isospin doublet, $I = \frac{1}{2}$.

The $\Xi$ is also called the cascade hyperon, or cascade particle, because it does not decay directly to the nucleon, but the decay is a cascade through the $\Lambda^0$,

$$\Xi^- \to \Lambda^0 + \pi^- \qquad \Delta S = +1$$
$$\Xi^0 \to \Lambda^0 + \pi^0 \qquad \Delta S = +1$$

followed by

$$\Lambda^0 \to p + \pi^- \qquad \Delta S = +1$$

or

$$\Lambda^0 \to n + \pi^0 \qquad \Delta S = +1$$

The hyperons all have antiparticles.

The strongly interacting particles as known in 1957 are summarized in Fig. 26.1 and Table 26.1.

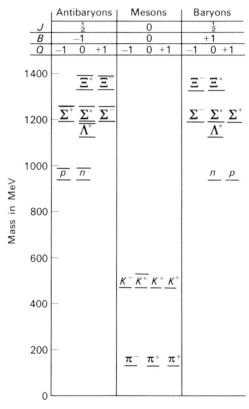

FIGURE 26.1    The strongly interacting particles known in 1957.

It should be noted that the strangeness is the same for each member of a given isospin multiplet.

Because of later developments in the theory of elementary particles, which will be dealt with in later chapters, it is now more usual to use hypercharge rather than strangeness. Hypercharge $Y$ is defined by

$$Y = B + S$$

Equation (24.5) then becomes

$$Q = I_3 + \tfrac{1}{2}Y$$

Since baryon number $B$ is conserved in all reactions, conservation or non-conservation of strangeness is equivalent to conservation or non-conservation respectively of hypercharge.

An interesting short history of discoveries in particle physics, including the discovery of strange particles, is given by Yang (1962).

TABLE 26.1 Quantum numbers of strongly interacting particles known in 1957

| | Baryons | | | | Mesons | | |
|---|---|---|---|---|---|---|---|
| | $S$ | $I$ | $I_3$ | | $S$ | $I$ | $I_3$ |
| $\Xi^-$ | $-2$ | $\frac{1}{2}$ | $-\frac{1}{2}$ | $K^0$ | $+1$ | $\frac{1}{2}$ | $-\frac{1}{2}$ |
| $\Xi^0$ | $-2$ | $\frac{1}{2}$ | $+\frac{1}{2}$ | $K^+$ | $+1$ | $\frac{1}{2}$ | $+\frac{1}{2}$ |
| $\overline{\Xi^-}$ | $+2$ | $\frac{1}{2}$ | $+\frac{1}{2}$ | $\overline{K^0}$ | $-1$ | $\frac{1}{2}$ | $+\frac{1}{2}$ |
| $\overline{\Xi^0}$ | $+2$ | $\frac{1}{2}$ | $-\frac{1}{2}$ | $K^-$ | $-1$ | $\frac{1}{2}$ | $-\frac{1}{2}$ |
| $\Sigma^-$ | $-1$ | $1$ | $-1$ | $\pi^-$ | $0$ | $1$ | $-1$ |
| $\Sigma^0$ | $-1$ | $1$ | $0$ | $\pi^0$ | $0$ | $1$ | $0$ |
| $\Sigma^+$ | $-1$ | $1$ | $+1$ | $\pi^+$ | $0$ | $1$ | $+1$ |
| $\overline{\Sigma^-}$ | $+1$ | $1$ | $+1$ | | | | |
| $\overline{\Sigma^0}$ | $+1$ | $1$ | $0$ | | | | |
| $\overline{\Sigma^+}$ | $+1$ | $1$ | $-1$ | | | | |
| $\Lambda^0$ | $-1$ | $0$ | $0$ | | | | |
| $\overline{\Lambda^0}$ | $+1$ | $0$ | $0$ | | | | |
| $n$ | $0$ | $\frac{1}{2}$ | $-\frac{1}{2}$ | | | | |
| $p$ | $0$ | $\frac{1}{2}$ | $+\frac{1}{2}$ | | | | |
| $\bar{n}$ | $0$ | $\frac{1}{2}$ | $+\frac{1}{2}$ | | | | |
| $\bar{p}$ | $0$ | $\frac{1}{2}$ | $-\frac{1}{2}$ | | | | |

### References

FOWLER, W. B., R. P. SHUTT, A. M. THORNDIKE and W. L. WHITTEMORE, *Phys. Rev.* **93** (1954) 861.

ROCHESTER, G. D. and C. C. BUTLER, *Nature* **160** (1947) 855.

YANG, C. N., *Elementary Particles*, 1962. Princeton University Press.

### Exercises

1  What was the experimental evidence leading to the idea of associated production of strange particles?

2    For a two-body decay

$$A \rightarrow B + C$$

show that in a frame of reference in which $A$ is at rest (the rest frame of $A$) the kinetic energy of particle $B$ is given by

$$T_B = \frac{\{(M_A - M_B)^2 - M_C^2\}\, c^2}{2M_A}$$

3    Using the result of Exercise 2, calculate the kinetic energy of each decay product for the following decays occurring at rest.

(a) $\pi^+ \rightarrow \mu^+ + v$
(b) $\pi^0 \rightarrow \gamma + \gamma$
(c) $K^+ \rightarrow \mu^+ + v$
(d) $K^+ \rightarrow \pi^+ + \pi^0$
(e) $\Lambda^0 \rightarrow p + \pi^-$
(f) $\Lambda^0 \rightarrow n + \pi^0$
(g) $\Sigma^+ \rightarrow p + \pi^0$
(h) $\Sigma^0 \rightarrow \Lambda^0 + \gamma$
(i) $\Xi^0 \rightarrow \Lambda^0 + \pi^0$
(j) $\Xi^+ \rightarrow \Lambda^0 + \pi^+$

4    Which of the following processes cannot occur through the strong interactions, and why?

(a) $K^- \rightarrow \pi^- + \pi^0$
(b) $K^+ \rightarrow \pi^+ + \pi^0 + \pi^0$
(c) $K^- + p \rightarrow \overline{K^0} + n$
(d) $\Sigma^+ + n \rightarrow \Sigma^- + p$
(e) $\Lambda^0 \rightarrow \Sigma^+ + \pi^-$
(f) $\Lambda^0 + n \rightarrow \Sigma^- + p$
(g) $\pi^+ + n \rightarrow K^+ + \Sigma^0$

5    Calculate the thresholds for the following reactions.

(a) $\pi^+ + n \rightarrow K^+ + \Sigma^0$
(b) $\pi^- + p \rightarrow K^0 + \Sigma^0$
(c) $p + p \rightarrow p + \Sigma^+ + K^0$

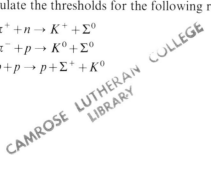

# Non-conservation of parity

# 7

## 27 The $\theta$–$\tau$ puzzle

A great deal of confusion occurred before physicists eventually arrived at the scheme of strange particles outlined in the previous chapter. We need not concern ourselves with most of that confusion, but we will look at one essential aspect which played an important part in shattering the complacency of physicists and led to the discovery of non-conservation of parity.

Amongst the decay modes of the $K$-meson, two different modes of decay are:

$$\theta \text{ mode}: K \rightarrow \pi + \pi \qquad (27.1)$$

$$\tau \text{ mode}: K \rightarrow \pi + \pi + \pi \qquad (27.2)$$

It was originally thought that there were in fact two different mesons, a $\theta$-meson and a $\tau$-meson, which decayed to two pions and three pions respectively. Experimentally it was found that the $\theta$ and $\tau$ particles had the same masses and life times. However, if one assumed that the decay reactions conserved parity, the $\theta$ and $\tau$ undoubtedly had different intrinsic parities.

Consider the decay of the $\theta^+$ at rest,

$$\theta^+ \rightarrow \pi^+ + \pi^0 \qquad (27.3)$$

Assuming that parity is conserved, we have

$$P_{\theta^+} = P_{\pi^+} P_{\pi^0} (-1)^l \qquad (27.4)$$

where $l$ is the orbital angular momentum of the $\pi^+ - \pi^0$ system. Since $P_{\pi^+} = P_{\pi^0} = -1$,

$$P_{\theta^+} = (-1)^l \qquad (27.5)$$

Since angular momentum must also be conserved and the pions have spin zero, we can deduce that $l = J$, where $J$ is the spin of the

$\theta^+$. So that

$$P_{\theta^+} = (-1)^J \qquad (27.6)$$

The spin and parity of the $\theta^+$ are restricted to

$$J^P = 0^+, 1^-, 2^+, \ldots \qquad (27.7)$$

Next we consider the decay of the $\tau^+$ at rest

$$\tau^+ \to \pi^+ + \pi^+ + \pi^- \qquad (27.8)$$

First consider the subsystem $(\pi^+ + \pi^+)$, which, consisting of two identical bosons, must have a wave function symmetric with respect to the interchange of the two pions, and so have even orbital angular momentum. So that the spin and parity of the two-pion subsystem are restricted to

$$J^P_{2\pi^+} = 0^+, 2^+, 4^+, \ldots \qquad (27.9)$$

The possible spins and parities of the system of three pions depends on the orbital angular momentum $l$ of the $\pi^-$. However, the experimental evidence rules out the possibility that the $\pi^-$ is emitted with non-zero orbital angular momentum $l$. For, if $l > 0$, fewer low energy $\pi^-$-mesons would be expected than are actually observed. Then $l = 0$, and since the $\pi^-$ has odd intrinsic parity, the spin and parity of the final system of three pions, and so, by conservation of angular momentum and parity, of the $\tau^+$-meson are restricted to

$$J^P = 0^-, 2^-, 4^-, \ldots \qquad (27.10)$$

These are not among the possible spin and parity assignments for the $\theta^+$, equation (27.7).

Either the $\theta^+$ and $\tau^+$ have different parities although they are identical in other respects, or alternatively parity is not conserved in the decays of $\theta$- and $\tau$-mesons.

The decay processes are weak interactions, with strengths comparable to the $\beta$-decay interaction. In 1956 Lee and Yang pointed out that there was no evidence for conservation of parity in $\beta$ decay or other weak interactions, although parity is conserved in strong interactions and electromagnetic interactions. Experiments on the $\beta$ decay of $Co^{60}$ were performed by Wu, Ambler, Hayward, Hoppes and Hudson in 1957 which demonstrated that parity was not conserved in $\beta$ decay. Experiments at Columbia University by Garwin, Lederman and Weinrich and at the University of Chicago by Friedman and Telegdi established that parity was not conserved in $\pi \to$

$\mu \rightarrow e$ decay. Very rapidly, many experiments were performed which demonstrated that parity was not conserved in the weak interactions.

The solution to the $\theta$–$\tau$ puzzle was that the $\theta$ and $\tau$ particles were the same, now called the $K$-meson, and the parity was not conserved in the decay of $K$-mesons.

## 28 Polarization of $\beta$ particles

The key experiment of Wu, Ambler and collaborators involved the radioactive $\beta^-$ emitter, $Co^{60}$, placed in a strong magnetic field and cooled so that the nuclei were aligned. It was a fairly complicated experiment. Later, easier ways of showing the non-conservation of parity were found. We will consider one of the simplest, the measurement of the longitudinal polarization of $\beta$ particles emitted by unpolarized nuclei.

To discuss the effect of conservation or non-conservation of parity on polarization, we need to know the effect that inversion of spatial coordinates through the origin has on spin. Figure 28.1 shows that for a classical spinning object, the direction of the angular momentum is unchanged by the inversion transformation. More generally, the effect of the inversion transformation on position and momentum is

$$\mathbf{r} \rightarrow -\mathbf{r}$$
$$\mathbf{p} \rightarrow -\mathbf{p}$$

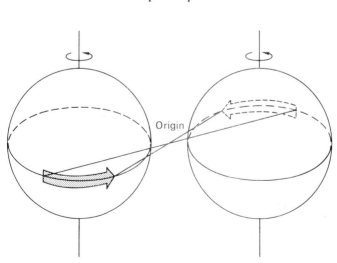

Origin

FIGURE 28.1   Effect of inversion on a classical spinning object.

so that the effect on angular momentum $\mathbf{M} = \mathbf{r} \times \mathbf{p}$ is

$$\mathbf{r} \times \mathbf{p} \rightarrow \mathbf{r} \times \mathbf{p}$$

Spin is also unaltered by the inversion transformation, as is shown quantum mechanically by Merzbacher (1970).

We consider an electron, with its spin pointing along its direction of motion, that has been emitted during the $\beta$ decay of a radioactive nucleus situated at the origin of our coordinate system, as shown in Fig. 28.2a. On performing an inversion through the origin on this system, the direction of motion of the electron is reversed, but the direction of its spin is unaltered, so that its spin is now pointing opposite to its direction of motion, as shown in Fig. 28.2b. If the physics of $\beta$ decay is to be invariant under inversion, the number of electrons emitted with spin parallel to the direction of motion must be the same as the number of electrons emitted with spin pointing opposite to the direction of motion.

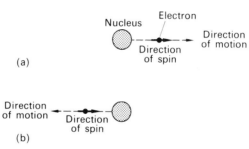

(a)

(b)

FIGURE 28.2    (a) Nucleus, at origin, emitting an electron with spin pointing along its direction of motion. (b) The effect of inversion; the spin is pointing opposite to the direction of motion.

Electrons emitted in the $\beta$ decay of $Co^{60}$ have been found to have a nett negative polarization along their direction of motion, i.e. more electrons are emitted with spins opposite the direction of motion than with spins parallel to the direction of motion. This polarization can be detected by first of all converting the longitudinally polarized beam into a transversely polarized one. An electric field is used to change the direction of motion by $90°$, the direction of the spin remaining unaltered in a non-relativistic approximation. See Fig. 28.3. Relativistic effects can be allowed for by deflecting the electrons through an angle slightly different from $90°$. The electric deflector also acts as a velocity selector, so that the subsequent polarization measurement is carried out on electrons with a certain velocity.

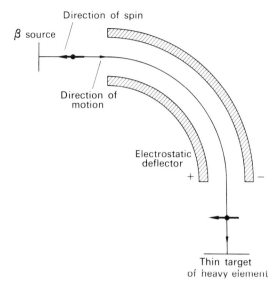

FIGURE 28.3   Measurement of longitudinal polarization of $\beta$ electrons. The longitudinal polarization is converted to transverse polarization by deflecting the electrons by an electric field. The electrons scattered by the target are detected in a plane perpendicular to the diagram.

The transverse polarization can be detected by measuring the scattering of the electrons by a thin target of a heavy element, and detecting any asymmetry in the scattering perpendicular to the direction of the spin of the electron. The electrons are scattered by the nuclei in the target. We consider the scattering by a nucleus of an electron with spin perpendicular to its direction of motion, as shown in Fig. 28.4. As well as the ordinary Coulomb force of attraction between the nucleus and the electron, there is also a spin–orbit force acting on the electron. This spin–orbit force is proportional to

$$\mathbf{s} . \mathbf{l} = \mathbf{s} . (\mathbf{r} \times \mathbf{p})$$

where $\mathbf{s}$ is the spin of the electron, and $\mathbf{l}$ is its orbital angular momentum. From Fig. 28.4 we see that the force acting on the electron differs according to whether the electron passes to the left or to the right of the plane through the scattering nucleus defined by the initial momentum and spin of the electron. The scattered beam will therefore be asymmetrical with respect to this plane.

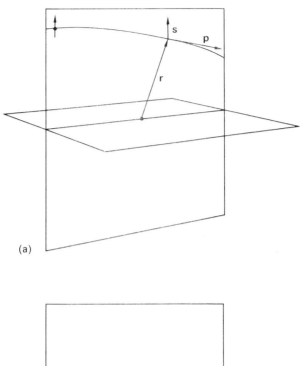

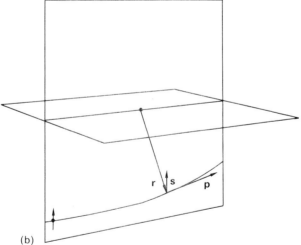

FIGURE 28.4    The scattering of an electron by the Coulomb field of a heavy nucleus. In (a) and (b), the spin **s** is perpendicular to the orbital angular momentum $\mathbf{l} = (\mathbf{r} \times \mathbf{p})$, and thus there is no spin–orbit force. In (c) and (d) the spin **s** is antiparallel and parallel, respectively, to the orbital angular momentum **l**, and thus the scattering in case (c) differs from that in case (d).

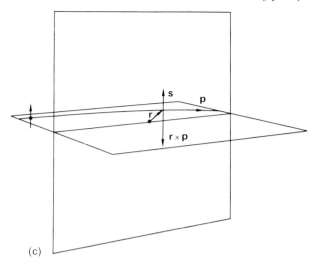

(c)

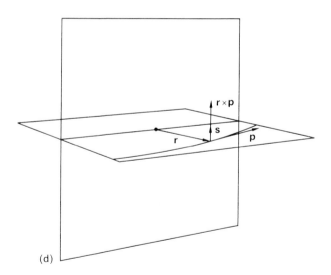

(d)

Therefore, any polarization of the initial electron beam can be detected by observing if there is any asymmetry in the scattered beam. A heavy element is used for the target, because the magnitude of the spin–orbit force increases with the atomic number of the scattering nucleus.

The longitudinal polarization of electrons emitted in the $\beta$ decay of $Co^{60}$ was measured by Frauenfelder *et al.* (1957) in the manner indicated by Fig. 28.3. They found that electrons with velocity $v = 0.49c$ had a longitudinal polarization of $-0.40$.

Subsequent experiments have shown that the longitudinal polarization of $\beta$ particles is $-v/c$ for electrons and $+v/c$ for positrons.

## 29  The two-component neutrino

In $\beta$ decay, a neutrino or antineutrino is emitted. The longitudinal polarization of electrons and the other phenomena of non-conservation of parity in $\beta$ decay can be explained by assuming that

(a) A neutrino always has spin pointing opposite to its direction of motion, called left-handed;

FIGURE 29.1.

(b) An antineutrino always has spin pointing parallel to its direction of motion, called right-handed.

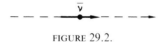

FIGURE 29.2.

A theory containing such neutrinos cannot conserve parity as it is not invariant under reflections, for a reflection would take a left-handed neutrino into a right-handed neutrino, and there are no right-handed neutrinos. The theory with left-handed neutrinos and right-handed antineutrinos is called the two-component neutrino theory.

In the Dirac theory of a particle with spin $\frac{1}{2}$, the wave function has four components. These four components are related to four possible states,

 (i)  the particle with spin up
 (ii)  the particle with spin down
(iii)  the antiparticle with spin up
(iv)  the antiparticle with spin down

where 'up' and 'down' refer to some chosen direction in space. Taking the chosen direction as the direction of motion, only states (ii) and (iii) occur for the neutrino, and correspondingly the neutrino can be described by a two-component wave function.

While it is possible to have a relativistically invariant theory in which only left-handed neutrinos occur, it is impossible to have a relativistically invariant theory in which only left-handed electrons occur. For, consider an electron with its spin pointing opposite to its direction of motion in some reference frame. In that reference frame, the electron is left-handed. In another reference frame, moving fast enough with respect to the first frame to overtake the electron, the direction of motion of the electron is reversed, and the electron appears right-handed.

However, a particle of zero rest mass, such as the neutrino, always travels at the speed of light, and so cannot be overtaken. Thus the handedness of a particle of zero rest mass has a relativistically invariant significance.

## 30 Non-conservation of parity in $\Lambda^0$ decay

Non-conservation of parity is a universal property of the weak interactions and is not restricted to processes involving neutrinos. Evidence that non-conservation of parity occurred for a decay in which no neutrino is involved was provided by experiments on the decay of the $\Lambda^0$ hyperon.

Consider the decay of a $\Lambda^0$ particle produced when a $\pi^-$ is incident on a proton

$$\pi^- + p \to \Lambda^0 + K^0 \tag{30.1}$$

Consider the decay mode

$$\Lambda^0 \to \pi^- + p \tag{30.2}$$

The direction of motion of the incident pion and the direction of motion of the $\Lambda^0$ together define a plane, as shown in Fig. 30.1. Let us call this plane the $\pi_{in}\Lambda^0$ plane. If parity is conserved, the probability of the $\pi^-$ from the decay of the $\Lambda^0$ being emitted to one side of the $\pi_{in}\Lambda^0$ plane should be the same as the probability of being emitted to the other side. For, if both production and decay of the $\Lambda^0$ are invariant under reflections, processes related by a reflection in the $\pi_{in}\Lambda^0$ plane, such as the examples in Fig. 30.1, must be equally probable. Experimentally, more $\pi^-$ from the $\Lambda^0$ decay are emitted to one side of the $\pi_{in}\Lambda^0$ plane than to the other.

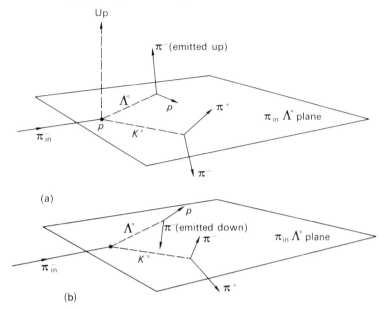

FIGURE 30.1    The production and decay of $\Lambda^0$. An incoming pion, $\pi_{in}^-$, collides with a proton $p$, producing a $\Lambda^0$ and a $K^0$. The dashed lines show the direction of the $\Lambda^0$ and $K^0$, which being uncharged are not observed directly, but determined from their charged decay products; (b) is the reflection of (a) in the $\pi_{in}\Lambda^0$ plane.

Let $\mathbf{p}_{\pi in}$, $\mathbf{p}_\Lambda$ and $\mathbf{p}_{\pi decay}$ be the momenta of the incident pion, the $\Lambda^0$ and the pion from the $\Lambda^0$ decay, respectively. We define 'up' relative to the $\pi_{in}\Lambda^0$ plane, by taking those pions from the $\Lambda^0$ decay as travelling up with respect to the $\pi_{in}\Lambda^0$ plane for which

$$(\mathbf{p}_{\pi in} \times \mathbf{p}_\Lambda) \cdot \mathbf{p}_{\pi decay} > 0$$

Similarly, a pion from the $\Lambda^0$ decay is classified as 'down' if

$$(\mathbf{p}_{\pi in} \times \mathbf{p}_\Lambda) \cdot \mathbf{p}_{\pi decay} < 0$$

Experimentally, more pions from $\Lambda^0$ decay are emitted upwards than downwards. For instance, from measurements with the kinetic energy of the incident pions ranging from 910 to 1300 MeV, Eisler *et al* (1957) obtained for the numbers of pions from the decay of $\Lambda^0$ particles emitted to each side,

$$\text{number 'up'} = 158$$
$$\text{number 'down'} = 105$$

clearly demonstrating parity violation in the decay of the $\Lambda^0$.

## 31 Invariance under *P*, *C* and *T*

Rotations, spatial translations and time displacements are continuous transformations, and, when they are symmetries of a system, are continuous symmetries. For instance, rotations can be performed through arbitrarily small angles.

As well as the continuous symmetries, there are discrete symmetries which are possible symmetries of physical systems. There are three fundamental discrete symmetry operations, namely the parity operation, time reversal and charge conjugation. We have already encountered the first of these, but include a brief account of it here with the other two for the convenience of listing together the properties of the discrete symmetries.

*Parity (P)*

This operator is associated with spatial inversion. For instance, for a single particle

$$P\psi(\mathbf{r}, t) = \xi\psi(-\mathbf{r}, t) \tag{31.1}$$

where $\xi$ is the intrinsic parity of the particle. For $n$ particles

$$P\psi(\mathbf{r}_1, \ldots, \mathbf{r}_n, t) = \xi_1 \ldots \xi_n \psi(-\mathbf{r}_1, \ldots, -\mathbf{r}_n, t) \tag{31.2}$$

where $\mathbf{r}_i$ is the position coordinate and $\xi_i$ the intrinsic parity of the $i^{\text{th}}$ particle. $P$ changes each $\mathbf{r}_i$ to $-\mathbf{r}_i$.

We have already shown (Section 13) that if

$$P\psi(\mathbf{r}) = \alpha\psi(\mathbf{r})$$

then $\alpha$ is necessarily $\pm 1$.

It was once assumed that all interactions were invariant under $P$. As a consequence of this assumption, if we choose $\psi(\mathbf{r}, 0)$ such that

$$P\psi(\mathbf{r}, 0) = \alpha\psi(\mathbf{r}, 0) \tag{31.3}$$

then

$$P\psi(\mathbf{r}, t) = \alpha\psi(\mathbf{r}, t) \tag{31.4}$$

Thus the parity would remain the same for all time, i.e. parity would be conserved. However, we know now that weak interactions are not invariant under $P$.

*Time reversal (T)*

$T$ is the operation of reversing all directions of motion including spin. The name that has been given to this operation, time reversal, is rather misleading; it would have been more appropriate to call it reversal of direction of motion. After all, time cannot be reversed, but directions of motion can be changed.

Invariance of interactions under $T$ means that if a certain physical process occurred, then the process obtained by reversing all directions of motion is also physically realizable.

Consider, for instance, the example in classical mechanics of a single planet moving in an elliptic orbit about a sun. Then motion in the same ellipse, but in the opposite direction, is also a physically realizable motion of the planet.

*Charge conjugation* $(C)$

$C$ is the operation of interchanging particles and antiparticles. Again we have a name which does not describe the operation exactly. For an electrically charged particle, the operation of charge conjugation does replace that particle by one of the opposite electric charge. But charge conjugation may also affect neutral particles. For instance $C$ replaces neutrons by antineutrons.

An important theorem of relativistic quantum mechanics is that, if it is in fact possible to describe all physical processes by means of relativistic field equations, then physics must be invariant under the combined operations of $PCT$, i.e. time reversal, charge conjugation and spatial inversion. This theorem is called the $PCT$ theorem. The order in which the operations of $P$, $C$ and $T$ are carried out has no effect. No violation of invariance under $PCT$ has ever been observed.

## 32 *CP* invariance

Figure 32.1a shows a schematic representation of the $\beta$ decay of $Co^{60}$

$$Co^{60} \rightarrow Ni^{60} + e^{-} + \bar{\nu} \tag{32.1}$$

in which a right-handed antineutrino is emitted. The charge conjugation $C$ of this decay, shown in Fig. 32.1b cannot occur as it involves a right-handed neutrino, and as discussed in Section 29 only left-handed neutrinos exist. Figure 32.1c shows the spatial inversion $P$ of the decay in Fig. 32.1a, and this also does not occur as left-handed antineutrinos do not exist. However, the process obtained by the combined operation of spatial inversion and charge conjugation, $CP$, on the $\beta$ decay of $Co^{60}$, shown schematically in Fig. 32.1d involves a right-handed neutrino and so can occur. Invariance under $CP$ would imply that the decay of anticobalt 60

$$\overline{Co^{60}} \rightarrow \overline{Ni^{60}} + e^{+} + \nu$$

has the same lifetime as the decay of cobalt 60, equation (32.1). No one has carried out experiments on anticobalt 60, which would be very difficult to prepare, but experiments on the decay of pions and muons show that the weak interactions are invariant under *CP*, at least to a very good approximation.

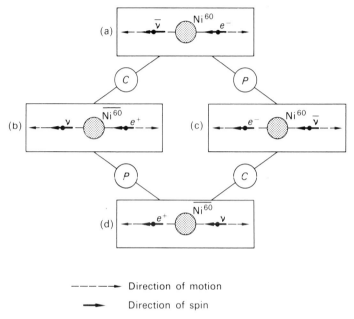

$$-----\blacktriangleright \quad \text{Direction of motion}$$
$$\blacktriangleright \quad \text{Direction of spin}$$

FIGURE 32.1    The effect of *C*, *P* and *CP* on the $\beta$ decay of Co$^{60}$.

Designating the left- and right-handed neutrinos by subscripts *L* and *R*, we summarize below the effect of *C*, *P* and *CP* for neutrinos according to the two-component neutrino theory.

$$P\nu_L = \nu_R, \text{ not physically realizable}$$
$$C\nu_L = \overline{\nu_L}, \text{ not physically realizable}$$
$$CP\nu_L = \overline{\nu_R}$$

Only $\nu_L$ and $\overline{\nu_R}$ are physically realizable.

In Section 28, we saw that, because of the lack of invariance under spatial inversion in $\beta$ decay, there is a built-in handedness in nature. However, because of the invariance of $\beta$ decay under *CP*, it is necessary to distinguish between matter and antimatter in order to use $\beta$ decay to determine a handedness.

Suppose we wish to explain to some person in outer space which is left and which is right. We could radio instructions to him on how to perform an experiment on the $\beta$ decay of $Co^{60}$, for instance the measurement of the polarization of the emitted electrons by observing the asymmetry of electron scattering as outlined in Section 28, and define 'left handedness' according to the scattering asymmetry. But, we cannot be sure that our hypothetical creature in outer space is not made of antimatter and so observe the positron decay of anticobalt 60. At present we believe that if he is made of antimatter, our instructions about left handedness would lead him to right handedness.

Observations of the decay of neutral $K$-mesons have shown that $CP$ is not absolutely conserved (see Section 39). However, for the present we can say that to a very large extent physics is invariant under $CP$, and therefore also invariant under $T$ because of the $PCT$ theorem. There is, as yet, no direct evidence for the failure of invariance under time reversal.

An important consequence of invariance under $PCT$ is that a particle and its antiparticle must have exactly the same mass and exactly the same lifetime.

## 33 Classification of interactions

The strengths of the various interactions between elementary particles enables a convenient and significant classification of these interactions.

The strength of the electromagnetic interaction is given by the fine structure constant, also called the electromagnetic coupling constant

$$e^2/\hbar c = 1/137 \cdot 036$$

Similarly, the strength of the other interactions can be given by appropriate coupling constants. Table 33.1 compares the strengths of the various interactions. The comparison of strengths of different types of interactions is only approximate, as the strength of an interaction can be defined in different ways. Table 33.1 shows the order of magnitude of the coupling constants for the various interactions.

The strong interactions are those between nucleons and other baryons, and $\pi$- and $K$-mesons in which strangeness is conserved. The particles participating in the strong interactions are called *hadrons*. The strong interactions also conserve isospin; the difference in properties of the different charge states of hadrons, such as the difference in mass between neutron and proton or the difference in mass between $\Sigma^0$ and $\Sigma^+$, can be attributed to the electromagnetic

TABLE 33.1    The strengths of elementary particle
interactions

| Interaction | Strength |
|---|---|
| Strong interactions (S.I.) | 1 |
| Electromagnetic interaction (E-M) | $10^{-2}$ |
| Weak interactions (W.I.) | $10^{-13}$ |
| Gravitational interaction | $10^{-38}$ |

interaction. (It should be noted however that we cannot yet calculate such mass differences.) Isospin and strangeness quantum numbers are assigned to the hadrons.

The electromagnetic interaction conserves strangeness but does not conserve isospin. The decay

$$\Sigma^0 \rightarrow \Lambda^0 + \gamma \qquad (33.1)$$

which conserves strangeness since both $\Sigma^0$ and $\Lambda^0$ have strangeness $-1$, occurs with a mean life of $< 10^{-14}$ s. Isospin is not conserved in the above decay since the $\Sigma^0$ has $I = 1$ and the $\Lambda^0$ has $I = 0$. Because of conservation of strangeness, the $\Lambda^0$ cannot decay to a neutron by emission of a photon,

$$\Lambda^0 \nrightarrow n + \gamma \qquad (33.2)$$

Since the electromagnetic interaction does not conserve isospin, the photon cannot be assigned a definite isospin. It is possible to regard the photon as having strangeness $S = 0$, but it is usual to assign isospin and strangeness only to hadrons. From equation (24.5), it is seen that the electromagnetic interaction conserves the third component of isospin $I_3$ since it conserves strangeness $S$. Baryon number $B$ and electric charge $Q$ are universally conserved.

Reactions which do not conserve strangeness proceed via the weak interactions. $\beta$ decay is also a weak interaction, as is also any interaction in which neutrinos are involved. We have seen that parity is not conserved in weak interactions, both in decays in which strangeness does not change such as $\beta$ decay, and in strangeness-changing decays such as the decay of the $K$-meson and the $\Lambda^0$.

The weakest observed interaction is the gravitational interaction. No effects of gravitational forces have yet been identified in elementary particle physics, so that the consideration of the gravitational interaction is generally omitted from discussions on elementary particles.

Table 33.2 summarizes the symmetries and conservation laws for the various interactions. The strong interactions have the maximum symmetry and the weak interactions the least. The ordering of interactions according to their symmetry is the same as the ordering according to their strength.

TABLE 33.2

| Symmetry | Space-time displace-ments | Spatial rota-tions | | | PCT | T | CP | C | P | | | |
|---|---|---|---|---|---|---|---|---|---|---|---|---|
| Conserved quantity | $E, \mathbf{p}$ | $\mathbf{J}$ | electric charge $Q$ | baryon number $B$ | | | | | $P$ | strangeness $S$ | 3rd component of isospin $I_3$ | isospin $I$ |
| 1. Strong interaction | ✓ | ✓ | ✓ | ✓ | ✓ | ✓ | ✓ | ✓ | ✓ | ✓ | ✓ | ✓ |
| 2. Electro-magnetic interaction | ✓ | ✓ | ✓ | ✓ | ✓ | ✓ | ✓ | ✓ | ✓ | ✓ | ✓ | × |
| 3 (i) weak inter-action | ✓ | ✓ | ✓ | ✓ | ✓ | ✓ | ✓ | × | × | × | × | × |
| (ii) Super-weak inter-action | ✓ | ✓ | ✓ | ✓ | ✓ | × | × | × | × | × | × | × |

✓ indicates that the interaction has that symmetry and that the appropriate quantity is conserved.

× indicates that the interaction violates that symmetry and that the appropriate quantity is not conserved.

For the superweak interaction, see the discussion of the decay of neutral $K$-mesons in Section 39.

## References

EISLER, F., R. PLANO, A. PRODELL, N. SAMIOS, M. SCHWARTZ, J. STEIN-BERGER, P. BASSI, V. BORELLI, G. PUPPI, G. TANAKA, P. WOLOSCHEK, V. ZOBOLI, M. CONVERSI, P. FRANZINI, I. MANNELLI, R. SANTANGELO, V. SILVESTRINI, D. A. GLASER, C. GRAVES, M. L. PERL, *Phys. Rev.* **108** (1957) 1353.

FRAUENFELDER, H., R. BOBONE, E. VON GOELER, N. LEVINE, H. R. LEWIS, R. N. PEACOCK, A. ROSSI and G. DE PASQUALI, *Phys. Rev.* **106** (1957) 386.

MERZBACHER, E., *Quantum Mechanics*, 2nd edition, 1970. Wiley, New York, p. 274.

### Exercises

1   In the experimental situation shown in Fig. 28.3, an electrostatic field is used to convert longitudinal polarization of electrons to transverse polarization. Can such conversion of polarization of electrons be achieved by deflection in a magnetic field? Can such conversion of polarization be achieved by deflection in a magnetic field for particles for which $g = 2$ exactly?

2   Classify the following processes as strong, electromagnetic, weak or totally forbidden and state whether the following are conserved in each process: parity, strangeness, isospin and third component of isospin.

   (a) $\Sigma^0 \rightarrow \Lambda^0 + \gamma$
   (b) $\Sigma^+ \rightarrow p + \pi^0$
   (c) $\Xi^- \rightarrow \Lambda^0 + \pi^-$
   (d) $\pi^- + p \rightarrow \Lambda^0 + K^0$
   (e) $\pi^- + p \rightarrow n + \pi^0$
   (f) $\gamma + p \rightarrow p + \pi^0$
   (g) $\Lambda^0 \rightarrow p + \pi^-$
   (h) $K^0 \rightarrow \pi^+ + \pi^-$.

3   Classify the following processes as strong, electromagnetic, weak or totally forbidden and state whether or not they conserve parity.

   (a) $\pi^0 \rightarrow \gamma + \gamma$
   (b) $K^+ \rightarrow \pi^+ + \pi^-$
   (c) $K^- \rightarrow \pi^- + \pi^- + \pi^+$
   (d) $\pi^+ \rightarrow \mu^+ + \nu$
   (e) $\mu^- \rightarrow e^- + \bar{\nu} + \nu$
   (f) $K^+ \rightarrow \mu^+ + \nu$
   (g) $\Sigma^+ \rightarrow \mu^+ + \nu$
   (h) $\Sigma^- \rightarrow n + e^- + \bar{\nu}$

# Leptons

8

### 34 Two kinds of neutrinos

Neutrinos are produced in the following two processes,

$$n \rightarrow p + e^- + \bar{\nu}_e$$
$$\pi^- \rightarrow \mu^- + \bar{\nu}_\mu$$

where the neutrinos are distinguished by subscripts because it has been found that the above neutrinos are not identical.

The difference between the two kinds of neutrinos was demonstrated at Brookhaven National Laboratory using the alternating gradient synchrotron, A.G.S. (Danby *et al.*, 1962; Lederman, 1963). 15 GeV protons within the machine collide with nuclei in a target. Among the particles produced are $\pi$-mesons which decay according to $\pi^+ \rightarrow \mu^+ + \nu_\mu$ and $\pi^- \rightarrow \mu^- + \bar{\nu}_\mu$. In the centre-of-mass frames of these decays, the products are emitted in all directions. And so, in the laboratory frame, the $\mu$-mesons and neutrinos move forward in a narrow cone. (See Exercises 2 and 3.) To remove all but the neutrinos from the beam, a wall of iron, 13·5 m thick, is placed in front of the detector. The Brookhaven A.G.S. was capable of accelerating protons to 30 GeV, but then 13·5 m of iron would have been insufficient to shield the detector from the particles other than neutrinos, and so the A.G.S. was run at 15 GeV. A plan view of the experiment is shown in Fig. 34.1.

The detector was a spark chamber with 90 aluminium plates, each about 1·2 m square, and with a total weight of about 10 tons.

In a spark chamber, the metal plates are arranged parallel to each other, as can be seen in Fig. 34·2, where the vertical white lines are light reflected from the edges of the aluminium plates. A potential difference is applied between neighbouring plates. A charged particle travelling through the spark chamber ionizes the gas in the gaps between the plates, and a spark occurs between the plates. The sparks between successive pairs of plates mark out the track traversed by the charged particle, as seen in Fig. 34.2.

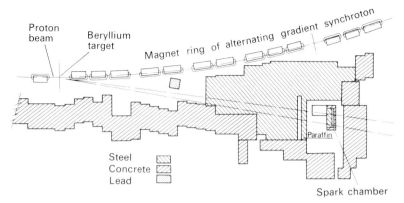

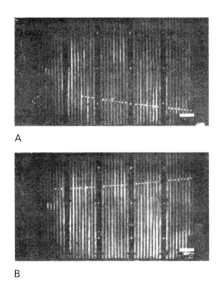

FIGURE 34.1   Plan view of the muon-neutrino experiment of Danby *et al.* (1962). Pions are produced by 15 BeV protons striking a beryllium target at one end of a 10-ft long straight section of the alternating-gradient synchrotron (AGS) at Brookhaven. (Only part of the AGS appears in the drawing.) About 10 per cent of the pions decay into muons and neutrinos before striking a 13·5 m thick iron shield wall at a distance of 21 m from the target. The shield wall stops the pions and muons but is easily penetrated by the neutrinos. Neutrino interactions are observed in a 10-ton aluminium spark chamber behind the shield wall. (After Danby *et al.*, 1962.)

FIGURE 34.2   Two examples of muon tracks observed in the experiment of Danby *et al.* (1962) on the muon neutrino.

In 300 h of observation, in which it was estimated that $10^{14}$ neutrinos passed through the spark chamber, 29 events of the type

$$\bar{v}_\mu + p \rightarrow n + \mu^+$$

were observed. Two examples of the tracks of $\mu$-mesons observed in the spark chamber are shown in Fig. 34.2. No events of the type

$$\bar{v}_\mu + p \rightarrow n + e^+$$

were observed. But we know from experiments on inverse $\beta$ decay, Section 9, that

$$\bar{v}_e + p \rightarrow n + e^+$$

is possible. We must conclude that $\bar{v}_\mu$ and $\bar{v}_e$ are different kinds of neutrinos; the neutrino associated with the $\mu$-meson is different from the neutrino associated with the electron.

## 35 The handedness of the muon neutrino

We have already seen in section 29 that neutrinos can have a definite relativistically invariant handedness because they have zero rest mass and so always travel with the velocity of light. In fact, the antineutrino $\bar{v}_e$ emitted along with an electron in $\beta$ decay is found to be right handed and the neutrino $v_e$ emitted along with a positron is found to be left handed.

Experiments on the non-conservation of parity in pion decay and muon decay show that the muon neutrino $v_\mu$ also has a definite handedness, and indeed that is left handed and so has the same handedness as $v_e$. Consider the decay

$$\pi^+ \rightarrow \mu^+ + v_\mu$$

in the rest frame of the initial pion and take the direction of the $z$ axis along the direction of motion of the decay products. To conserve angular momentum the $z$ component of the spins of $\mu^+$ and $v_\mu$ must be equal and of opposite sign because the pion has zero spin. $\mu^+$ and $v_\mu$ are emitted in opposite directions and so, in the rest frame of the initial pion, $\mu^+$ and $v_\mu$ have the same handedness. See Fig. 35.1. The handedness of the $v_\mu$ can be obtained by measuring the $z$ component of the spin of the emitted $\mu^+$. Measurement of the

FIGURE 35.1    Decay of a positive pion.

longitudinal polarization of muons showed that $v_\mu$ is left handed and $\bar{v}_\mu$ is right handed.

## 36 Conservation of leptons

Electrons, muons and neutrinos and their antiparticles are called leptons. Before it was known that $v_\mu$ and $v_e$ were different, a law of conservation of lepton number was postulated for all interactions, and $e^-$, $\mu^-$ and $v$ were assigned a lepton number $+1$, and $e^+$, $\mu^+$ and $\bar{v}$ were assigned a lepton number $-1$. However, with this scheme, such reactions as

$$\bar{v}_\mu + p \rightarrow n + e^+$$

would not be forbidden on the grounds of conservation of lepton number, and as discussed in Section 34, such a reaction is not observed. To remedy this, two different lepton numbers were introduced, a muon-lepton number $L_\mu$, and an electron-lepton number $L_e$, and it is required that these two lepton numbers be separately conserved in all interactions. The muon lepton numbers and electron lepton numbers are shown in Table 36.1. For all other particles, the hadrons or strongly interacting particles and the photon, the electron-lepton number $L_e$ and the muon-lepton number $L_\mu$ are both zero.

TABLE 36.1

|  | $e^-$ | $e^+$ | $v_e$ | $\bar{v}_e$ | $\mu^-$ | $\mu^+$ | $v_\mu$ | $\bar{v}_\mu$ |
|---|---|---|---|---|---|---|---|---|
| $L_e$ | $+1$ | $-1$ | $+1$ | $-1$ | $0$ | $0$ | $0$ | $0$ |
| $L_\mu$ | $0$ | $0$ | $0$ | $0$ | $+1$ | $-1$ | $+1$ | $-1$ |

As a consequence of the conservation of $L_\mu$ and $L_e$, if certain leptons disappear in a reaction then certain leptons are always produced. Thus in drawing Feynman diagrams for the weak interactions, a particular lepton line can join on to another lepton line according to the following

$$e^- \leftrightarrow v_e$$
$$e^+ \leftrightarrow \bar{v}_e$$
$$\mu^- \leftrightarrow v_\mu$$
$$\mu^+ \leftrightarrow \bar{v}_\mu$$

(Of course, other conservation laws such as conservation of charge must be satisfied.) It should be remembered that reversing the direction of the arrow on a line in a Feynman diagram means replacing the particle by its antiparticle. Some examples of weak interactions are given below. The lepton numbers $L_e$ and $L_\mu$, and the baryon number $B$ are shown for some of the reactions.

(a)

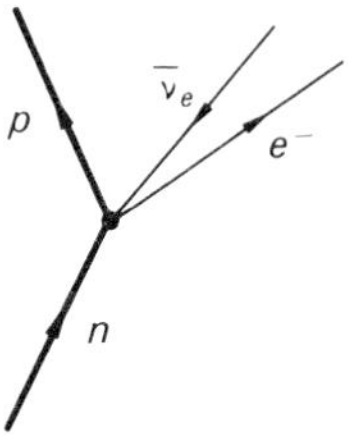

$$n \to p \quad + \bar{\nu}_e \quad + e^-$$
$$B \quad +1 = +1+0 \quad +0$$
$$L_e \quad 0 = 0 \quad +(-1)+1$$

FIGURE 36.1.

(b)

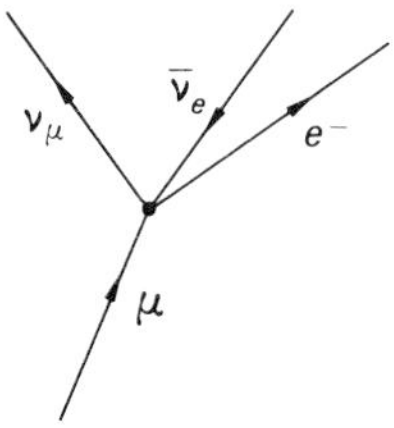

$$\mu^- \to e^- + \bar{\nu}_e \quad + \nu_\mu$$
$$B \quad 0 = 0 \ +0 \quad +0$$
$$L_e \quad 0 = 1 \ +(-1)+0$$
$$L_\mu \quad 1 = 0 \ +0 \quad +1$$

FIGURE 36.2.

(c)

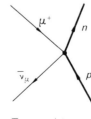

$$p + \bar{\nu}_\mu \to \mu^+ + n$$

FIGURE 36.3.

In the Feynman diagram of a weak interaction, four fermion lines meet at a vertex, as seen in the diagrams for examples (a), (b) and (c) above. For some reactions or decays proceeding by the weak interaction, the appropriate Feynman diagram contains other vertices involving the strong interactions as well as the weak interaction vertex with four fermion lines. The lowest order Feynman diagrams for pion decay are shown in Fig. 36.4. For pion decay, further diagrams with more strong interaction vertices will be important. Because the coupling constant of the weak interaction is so small, diagrams with additional weak interaction vertices can be neglected.

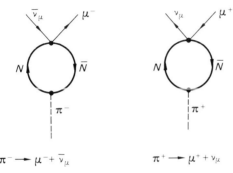

FIGURE 36.4    Pion decay.

The weak interaction is a coupling between four fermions and is called a four-fermion interaction.

The Feynman diagrams in Fig. 36.4 show the decay of the pion proceeding by creation of a virtual nucleon–antinucleon pair which in turn gives rise to a muon and a neutrino by means of the four-fermion interaction.

## 37  Universal conservation laws

There are certain conservation laws for which no violation has ever been observed. Thus the following quantities are always conserved in all interactions

1. Energy
2. Momentum $\Big\}$ related to
3. Angular momentum $\Big\}$ space-time symmetries
4. Charge
5. Electron lepton number

6. Muon lepton number
7. Baryon number
8. Number of fermions minus number of antifermions.

Any reaction or decay not forbidden by the above conservation laws occurs, although in some cases, only in amounts difficult to observe because other reactions or decay are far more probable. As examples, consider the variety of ways in which the $K^+$ and the $\Sigma^+$ can decay, shown in Table 37.1.

TABLE 37.1

| Partial decay mode | Fraction of total decays |
|---|---|
| $K^+ \rightarrow \mu^+ + \nu_\mu$ | 64% |
| $\pi^+ + \pi^0$ | 21% |
| $\pi^+ + \pi^- + \pi^+$ | 5·6% |
| $\pi^+ + \pi^0 + \pi^0$ | 1·7% |
| $\mu^+ + \pi^0 + \nu_\mu$ | 3·2% |
| $e^+ + \pi^0 + \nu_e$ | 4·9% |
| $\pi^+ + \pi^- + e^+ + \nu_e$ | $3\cdot3 \times 10^{-5}$ |
| $\pi^+ + \pi^+ + e^- + \bar{\nu}_e$ | $<7 \times 10^{-7}$ |
| $\pi^+ + \pi^- + \mu^+ + \nu_\mu$ | $0\cdot9 \times 10^{-5}$ |
| $\pi^+ + \pi^+ + \mu^- + \bar{\nu}_\mu$ | $<3 \times 10^{-6}$ |
| $e^+ + \nu_e$ | $1\cdot2 \times 10^{-5}$ |
| $\pi^+ + \mu^+ + \mu^-$ | rare |
| $\pi^+ + e^+ + e^-$ | rare |
| any of above $+ \gamma$ | rare |
| $\Sigma^+ \rightarrow p + \pi^0$ | 52% |
| $n + \pi^+$ | 48% |
| $p + \gamma$ | $1\cdot2 \times 10^{-3}$ |
| $n + \pi^+ + \gamma$ | $1\cdot3 \times 10^{-4}$ |
| $\Lambda^0 + e^+ + \nu_e$ | $2 \times 10^{-5}$ |
| $n + \mu^+ + \nu_\mu$ | $<2\cdot3 \times 10^{-5}$ |
| $n + e^+ + \nu_e$ | $<1\cdot1 \times 10^{-5}$ |

Information for this table is from *Phys. Letters* **33B** (1970) No. 1

In addition to the above conservation laws, which hold for all interactions, we have that

1. Weak interactions are invariant under $PC$ and $T$ (at least to a good approximation. See Section 39 concerning non-conservation of $PC$).

2. Electromagnetic interactions are invariant under $P$, $C$ and $T$ separately and also conserve strangeness $S$ and the third component of isospin $I_3$.

3. Strong interactions are invariant under $P$, $C$ and $T$ separately and conserve strangeness and isospin (for the strong interactions, both $I$ and $I_3$ are good quantum numbers).

### References

DANBY, G., J-M. GAILLARD, K. GOULIANOS, L. M. LEDERMAN, N. MISTRY, M. SCHWARTZ and J. STEINBERGER, *Phys. Rev. Letters* **9** (1962) 36.

L. M. LEDERMAN, 'The two-neutrino experiment', *Sci. Amer.*, March 1963, p. 60.

### Exercises

1 For the following decays, label the neutrinos according to whether they are electron-neutrinos or muon-neutrinos, neutrinos or antineutrinos.
   (a) $\Lambda^0 \rightarrow p + \mu^- + \nu$
   (b) $\Sigma^+ \rightarrow \Lambda^0 + e^+ + \nu$
   (c) $\Sigma^- \rightarrow n + e^- + \nu$
   (d) $\mu^+ \rightarrow e^+ + \nu + \nu$
   (e) $K^0 \rightarrow \pi^- + e^+ + \nu$

2 Find the angular distribution in the laboratory frame of neutrinos from the decay of moving pions.

3 For the case of pions with kinetic energy of 3 BeV, graph the angular distribution of neutrinos using the result of the previous exercise. Also calculate the maximum angle in the laboratory frame between the direction of the decay muons and the pions.

4 Write down all possible decays of the $\Lambda^0$ allowed by the universal conservation laws. Check from the *Review of Particle Properties* which decays have been observed. (See end of Section 41 (p. 99) concerning *Review of Particle Properties*.)

# Neutral $K$-mesons and non-conservation of $CP$

**9**

## 38 Neutral $K$-mesons

As we saw in Section 25, there are two kinds of neutral $K$-mesons, the $K^0$ and the $\overline{K^0}$. Each is the antiparticle of the other. The $K^0$ has strangeness $S = +1$, and the $\overline{K^0}$ has $S = -1$. So certain reactions proceed strongly for one kind of neutral $K$-meson and not for the other. For instance

$$
\begin{aligned}
K^0 + p &\to K^+ + n & \overline{K^0} + p &\not\to K^+ + n \\
K^0 + n &\not\to K^- + p & \overline{K^0} + n &\to K^- + p \\
K^0 + n &\not\to \pi^+ + \Sigma^- & \overline{K^0} + n &\to \pi^+ + \Sigma^-
\end{aligned}
\tag{38.1}
$$

and $K^0$ and $\overline{K^0}$ are produced in different reactions, e.g.

$$
\begin{aligned}
p + n &\to p + \Lambda^0 + K^0 \\
S = 0 + 0 &= 0 \quad -1 + 1
\end{aligned}
\tag{38.2}
$$

$$
\begin{aligned}
K^- + p &\to \overline{K^0} + n \\
S = -1 + 0 &\doteq -1 + 0
\end{aligned}
\tag{38.3}
$$

We now consider the action of charge conjugation $C$ and inversion of spatial coordinates (parity operation) $P$ on neutral $K$-mesons. Let $\psi_{K^0}$ and $\psi_{\overline{K^0}}$ be the state functions describing a $K^0$ and a $\overline{K^0}$ respectively, at rest. Since the kaon has odd intrinsic parity

$$
\begin{aligned}
P\psi_{K^0} &= -\psi_{K^0} \\
P\psi_{\overline{K^0}} &= -\psi_{\overline{K^0}}
\end{aligned}
\tag{38.4}
$$

It may be noted that there is a certain arbitrariness in assigning odd intrinsic parity to the kaon. The parity of $S = 1$ particles cannot be absolutely determined in relation to $S = 0$ particles, for the only interactions connecting states of different strangeness are weak interactions and do not conserve parity. Once an intrinsic parity is assigned to any $S = 1$ particle, the intrinsic parities of all other strange particles can be determined. We could assign even intrinsic parity to the $K$-meson, but then the $S = -1$ baryons would have odd intrinsic parity. It is more convenient to use a scheme in which the $\Sigma$, $\Lambda$ and $\Xi$ have the same intrinsic parity as the nucleon, and kaons have the same intrinsic parity as pions.

Charge conjugation $C$ is the replacement of particle by anti-particle and so takes $|\psi_{K^0}|^2$ into $|\psi_{\overline{K^0}}|^2$,

$$|C\psi_{K^0}|^2 = |\psi_{\overline{K^0}}|^2$$
$$|C\psi_{\overline{K^0}}|^2 = |\psi_{K^0}|^2$$

so that

$$C\psi_{K^0} = \eta\psi_{\overline{K^0}} \tag{38.5}$$

where $|\eta| = 1$. The relative phase of $\psi_{K^0}$ and $\psi_{\overline{K^0}}$ is usually chosen so that

$$CP\psi_{K^0} = \psi_{\overline{K^0}} \tag{38.6}$$
$$CP\psi_{\overline{K^0}} = \psi_{K^0}$$

$\psi_{K^0}$ and $\psi_{\overline{K^0}}$ are not eigenstates of $CP$, but the states

$$\psi_{K_1^0} = \frac{1}{\sqrt{2}}(\psi_{K^0} + \psi_{\overline{K^0}}) \tag{38.7}$$

$$\psi_{K_2^0} = \frac{1}{\sqrt{2}}(\psi_{K^0} - \psi_{\overline{K^0}}) \tag{38.8}$$

are (normalized) eigenstates of $CP$, for

$$CP\psi_{K_1^0} = +\psi_{K_1^0} \tag{38.9}$$

$$CP\psi_{K_2^0} = -\psi_{K_2^0} \tag{38.10}$$

Because of conservation of $CP$ in the weak interactions, the $K_1^0$ cannot decay into states with $CP = -1$ and the $K_2^0$ cannot decay into states with $CP = +1$, so that the $K_1^0$ and the $K_2^0$ have different modes of decay.

For instance, consider the state of two neutral $\pi$-mesons in the centre-of-mass frame. Since the product of the intrinsic parities of the pions is $+1$, the effect of the parity operator is simply to interchange the two pions. But this interchange must leave the wave function unaltered because the $\pi^0$-mesons are identical bosons. Writing the wave function of the two pions as $\phi(\pi^0, \pi^0)$, then

$$P\phi(\pi^0, \pi^0) = +\phi(\pi^0, \pi^0) \tag{38.11}$$

Furthermore, since the $\pi^0$ is its own antiparticle

$$CP\phi(\pi^0, \pi^0) = +\phi(\pi^0, \pi^0) \qquad (38.12)$$

Next consider the system of a $\pi^+$ and a $\pi^-$ in the centre-of-mass frame. Again the effect of the parity operator is to interchange the two particles, i.e.

$$P\phi(\pi^+, \pi^-) = +\phi(\pi^-, \pi^+) \qquad (38.13)$$

Since $\pi^+$ and $\pi^-$ are antiparticles of each other, it follows that

$$CP\phi(\pi^+, \pi^-) = +\phi(\pi^+, \pi^-) \qquad (38.14)$$

Since the system of two pions has $CP = +1$, only the $K_1^0$ can decay into two pions.

$$K_1^0 \rightarrow \pi^0 + \pi^0$$
$$K_1^0 \rightarrow \pi^+ + \pi^-$$

The $K_1^0$ has a mean life of $0.86 \times 10^{-10}$ s.

The $K_2^0$ decays by other modes, such as

$$
\begin{aligned}
K_2^0 &\rightarrow \pi^0 + \pi^0 + \pi^0 \\
&\rightarrow \pi^+ + \pi^- + \pi^0 \\
&\rightarrow \pi + \mu + \nu \\
&\rightarrow \pi + e + \nu
\end{aligned}
\qquad (38.15)
$$

and has a mean life of $5 \times 10^{-8}$ s.

As the lifetimes of the $K_1^0$ and $K_2^0$ are so different, we are led to consider the $K_1^0$ and $K_2^0$ as particles, rather than $K^0$ and $\overline{K^0}$. However, neutral kaons are produced as either $K^0$ or $\overline{K^0}$, and then subsequently decay as $K_1^0$ or $K_2^0$. For instance, consider a beam of $K^0$ particles produced by reaction (38.2). Since from equations (38.7) and (38.8)

$$\psi_{K^0} = \frac{1}{\sqrt{2}} (\psi_{K_2^0} + \psi_{K_1^0}) \qquad (38.16)$$

half of the beam decays quickly as $K_1^0$, the other half taking longer to decay as $K_2^0$. See Fig. 38.1a. Thus at some distance along a beam of neutral $K$-mesons, virtually all of the $K_1^0$ part of the beam will have disappeared because of the relatively rapid $2\pi$ decay of the $K_1^0$. The beam will then mainly consist of $K_2^0$ particles. We shall call this the 'stale' beam. However, we see from equation (38.8), that if we allow the 'stale' beam to undergo some strong interaction we will be able to detect $K^0$-mesons in the beam. For instance, the 'stale' beam can be used to produce any of the reactions in equation (38.1).

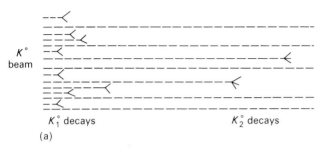

$K_1^\circ$ decays            $K_2^\circ$ decays

(a)

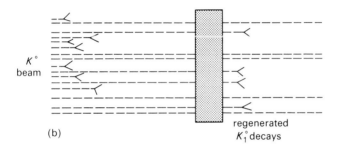

regenerated
$K_1^\circ$ decays

(b)

 decay into $2\pi$

decay into $3\pi$

FIGURE 38.1    Decay of neutral kaons.

If the 'stale' beam passes through a target, the $K^0$ and $\overline{K^0}$ parts of the $K_2^0$ beam will interact differently with the target nuclei because of their different strangeness. As a result, when the beam emerges from the target, the relative amplitudes and phases of the $K^0$ and $\overline{K^0}$ states which contribute to the beam will have changed, and the beam will consist of particles in the state

$$\psi = \alpha\psi_{K^0} - \beta\psi_{\overline{K^0}}$$

with $\alpha \neq \beta \neq 1/\sqrt{2}$. The beam is no longer a pure $K_2^0$ beam; $K_1^0$ particles reappear, regenerated by the target. After the beam has passed through the target, decays into two pions are observed. See Fig. 38.1b.

Neutral $K$-mesons could be described in terms of any two ortho-gonal states obtained from linear superposition of $K^0$ and $\overline{K^0}$; but this generally would not be convenient. It is simplest to describe neutral kaons in terms of states of definite strangeness, the $K^0$ and $\overline{K^0}$, when treating strong interactions since strangeness is conserved in the strong interactions, and in terms of states of definite $CP$, $K_1^0$ and $K_2^0$, when treating weak interactions which conserve $CP$.

The description of the behaviour of neutral kaons in terms of either $K^0$ and $\overline{K^0}$ on the one hand or $K_1^0$ and $K_2^0$ on the other, has a close analogy with the description of light in terms of right-handed and left-handed circularly polarized light on the one hand, or in terms of plane polarized light on the other. Plane polarized light can be described as a linear superposition of right-handed and left-handed circularly polarized light. Consider a beam of right-handed circularly polarized light travelling along the $z$ direction incident on a filter which transmits only light plane-polarized in the $x$ direction. If this now plane-polarized light passes through a second filter with different transmission properties for right-handed and left-handed circularly polarized light, the final beam of light is no longer plane-polarized in the $x$ direction but contains also a component plane-polarized in the $y$ direction which has been regenerated by the second filter in an analogous manner to the regeneration of $K_1^0$-mesons.

## 39  Non-conservation of *CP*

In the previous section, we saw that assuming that the weak inter-actions are invariant under $CP$, a beam of $K^0$-mesons consists of a short-lived component $K_1^0 \rightarrow \pi + \pi$ and a long-lived component $K_2^0 \rightarrow \pi + \pi + \pi$. So that at any considerable distance, much greater than $c\tau = 2 \cdot 59$ cm, where $\tau$ is the mean life of the $K_1^0$, from the source of neutral kaons, no decay into two pions should be observed.

In 1964, Christenson, Cronin, Fitch and Turlay observed decays into two charged pions in a beam of neutral kaons at a distance of 57 ft from where the kaons were produced. As no $K_1^0$-mesons could have survived to this distance, they had observed the decay

$$K_2^0 \rightarrow \pi^+ + \pi^- \tag{39.1}$$

Since, as discussed in the previous section, the $K_2^0$ has $CP = -1$, and the system of two pions has $CP = +1$, they had observed the non-conservation of $CP$ and so shown that the weak interactions are not invariant under the combined operation of spatial inversion and charge conjugation.

The detector arrangement of the experiment is shown in Fig. 39.1. The decay of a $K_2^0$ beam in helium gas was observed using spark chambers triggered by Cerenkov counters in coincidence. Two coincident charged particles were detected and their momenta

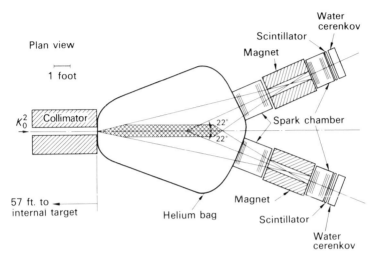

FIGURE 39.1   Plan view of the detector arrangement of Christenson *et al.* (1964). The volume from which decays were observed is cross-hatched.

measured. Assuming that the particles observed were π-mesons, the invariant mass $M^*$ was calculated, i.e.

$$M^* = c^{-2}\{(E_1 + E_2)^2 + c^2(\mathbf{p}_1 + \mathbf{p}_2)^2\}^{\frac{1}{2}} \tag{39.2}$$

using

$$E_i = (c^2 p_i^2 + M_\pi^2 c^4)^{\frac{1}{2}} \tag{39.3}$$

$M^*$ corresponds to the rest mass of the decaying particle if the decay was only into 2π. For the decay, equation (39.1),

$$M^* = M_K = 498 \text{ MeV} \tag{39.4}$$

For

$$K_2^0 \rightarrow \pi^+ + \pi^- + \pi^0 \tag{39.5}$$

only the charged pions are observed, and

$$280 \text{ MeV} < M^* < 363 \text{ MeV} \tag{39.6}$$

and there can be no confusion with the decay of equation (39.1).

For

$$K_2^0 \rightarrow \pi + \mu + \nu \qquad (39.7)$$

$$280 \text{ MeV} < M^* < 516 \text{ MeV} \qquad (39.8)$$

and for

$$K_2^0 \rightarrow \pi + e + \nu \qquad (39.9)$$

$$280 \text{ MeV} < M^* < 536 \text{ MeV} \qquad (39.10)$$

For both reactions (39.7) and (39.9), $M^*$ would vary smoothly over the ranges given, and there is no reason for these decays to have their values of $M^*$ peaked near 498 MeV.

The angle $\theta$ between the direction of the $K_2^0$ beam and the vector sum of the momenta of the two observed particles was also determined. For two-body decays this angle should be zero, and, in general, is different from zero for three-body decays. $\theta$ was found to be close to zero for events with $M^* \simeq M_K$.

The observations were consistent with the assumption of $K_2^0$ decaying into two $\pi$-mesons.

The possible reaction

$$K_2^0 \rightarrow \pi^+ + \pi^- + \gamma \qquad (39.11)$$

could only have produced the observed results if the $\gamma$ ray energy was restricted to values less than 1 MeV out of the available kinetic energy of up to 209 MeV.

Out of a sample of 22 700 $K_2^0$ decays, $45 \pm 9$ were identified as being $K_2^0 \rightarrow \pi^+ + \pi^-$. This number was too large to be explained on the basis of regeneration of $K_1^0$ in the helium gas or by regeneration elsewhere.

The apparatus was calibrated by observing the decays of $K_1^0 \rightarrow 2\pi$ when $K_1^0$-mesons were produced by regeneration by placing a tungsten target in the beam.

The experiment on the decay $K_2^0 \rightarrow \pi^+ + \pi^-$ has since been repeated by several groups with the same result. Various suggestions were advanced to try to save $CP$ invariance, but none of them has proved consistent with the experimental results, and so we must accept the conclusion that $CP$ is not conserved. Hence by the $CPT$ theorem, we must accept that invariance under time reversal, $T$, also does not hold.

In one respect, the non-conservation of $CP$ is less satisfactory aesthetically than the non-conservation of parity. When it was discovered that parity conservation was violated in weak interactions, it was at least some consolation that it was maximally violated, in the sense that all the antineutrinos produced were righthanded, and not just something like 51 per cent of them. But the non-conservation of $CP$ is a small effect, and $CP$ is very nearly conserved. For instance, the result of the experiment of Christenson *et al.* was a branching ratio

$$R = (K_2^0 \rightarrow \pi^+ + \pi^-)/(K_2^0 \rightarrow \text{all charged modes})$$
$$= (2 \cdot 0 \pm 0 \cdot 4) \times 10^{-3}$$

Since the weak interaction causing the decay of the neutral kaon does not conserve $CP$, the short-lived neutral kaon and the long-lived neutral kaon are not necessarily eigenstates of $CP$. However, it is now conventional to use $K_1^0$ and $K_2^0$ for the eigenstates of $CP$, as given by equations (37.7) and (37.8), and to use $K_L$ for the long-lived neutral kaon and $K_S$ for the short-lived neutral kaon. With this notation, the decay observed by Christenson *et al.* is described as

$$K_L \rightarrow \pi^+ + \pi^- \tag{39.12}$$

A detailed account of non-conservation of $CP$ and the decays of neutral kaons is given by Kabir (1968). The recent status of $CP$ violation has been reviewed by Steinberger (1969).

As well as the decay of equation (39.12),

$$K_L \rightarrow \pi^0 + \pi^0 \tag{39.13}$$

has also been measured. The results are given by the parameters

$$\eta_{+-} = \frac{\text{amplitude } (K_L \rightarrow \pi^+ + \pi^-)}{\text{amplitude } (K_S \rightarrow \pi^+ + \pi^-)} \tag{39.14}$$

$$\eta_{00} = \frac{\text{amplitude } (K_L \rightarrow \pi^0 + \pi^0)}{\text{amplitude } (K_S \rightarrow \pi^0 + \pi^0)} \tag{39.15}$$

The results are shown in Table 39.1. Note that equations (39.14) and (39.15) are ratios of quantum mechanical amplitudes, and the ratio of number of events, or probability ratio, is given by $|\eta_{+-}|^2$ and $|\eta_{00}|^2$.

TABLE 39.1.   Experimental results on $CP$ violation in neutral kaon decay (From Review of Particle Properties, Söding *et al. Phys. Letters* **39B** (1972) No. 1.)

| |
|---|
| $\eta_{+-} = (1 \cdot 96 \pm 0 \cdot 03) \times 10^{-3} \exp[i(43 \pm 3)^\circ]$ |
| $\eta_{00} = (2 \cdot 09 \pm 0 \cdot 12) \times 10^{-3} \exp[i(43 \pm 19)^\circ]$ |

The violation of $CP$ has also been seen in the decays

$$K_L \to \pi^+ + \mu^- + \bar{\nu}_\mu \qquad (39.16)$$

$$K_L \to \pi^- + \mu^+ + \nu_\mu \qquad (39.17)$$

$$K_L \to \pi^+ + e^- + \bar{\nu}_e \qquad (39.18)$$

$$K_L \to \pi^- + e^+ + \nu_e \qquad (39.19)$$

where a small charge asymmetry is observed, i.e. the number of decays (39.16) and (39.17) is not the same, and similarly the number of decays (39.18) and (39.19) is not the same.

### References

CHRISTENSON, J. H., J. W. CRONIN, V. L. FITCH and R. TURLAY, *Phys. Rev. Letters* **13** (1964) 138.

KABIR, P. K., *The CP Puzzle: Strange Decays of the Neutral Kaon,* 1968. Academic Press, London and New York.

ROOS, M., C. BRICMAN, A. BARBARO-GALTIERI, L. R. PRICE, A. RITTEN-BERG, A. H. ROSENFELD, N. BARASH-SCHMIDT, P. SÖDING, C. Y. CHIEN, C. G. WOHL, T. LASINSKI, *Phys. Letters* **33B** (1970) 1.

STEINBERGER, J., *Comments on Nuclear and Particle Physics* **3** (1969) 73; *Proceedings of the Lund International Conference on Elementary Particles, 1969.*

### Exercises

1   For the following decays, assign all possible electric charges to the decay particles, and identify the neutrinos, i.e. $\nu_\mu, \bar{\nu}_\mu, \nu_e, \bar{\nu}_e$. Give the spin of each particle and state whether each particle is a boson or fermion. Calculate $Q$, the energy released in the decay
   (a) $K_S^0 \to \pi + \pi$
   (b) $K_L^0 \to \pi + \pi + \pi$
   (c) $K_L^0 \to \pi + \mu + \nu$
   (d) $K_L^0 \to \pi + e + \nu$.

2   Find the ratio of $K_S$ to $K_L$ in a beam of 10 GeV neutral kaons at a distance of 20 m from where the beam is produced.

# Resonances

**10**

## 40 Introduction

We saw in Section 24 that the long lifetime of strange particles was explained by conservation of strangeness; the strange particles were produced by strong interactions and subsequently decayed by weak interactions. What about particles which decay by strong interactions? Such particles have a very brief existence, with lifetimes corresponding to the time for a fast particle to travel a distance of the range of nuclear forces, i.e. $\tau \simeq 10^{-23}$ s, and so strongly decaying particles cannot travel far enough to leave tracks in a cloud chamber or bubble chamber, or be directly observed in any way.

Since strongly decaying particles have such an ephemeral existence, we wonder whether to call them particles at all, but physicists have found it convenient to extend their use of the word 'particle' to include such objects, following no doubt the example of Humpty Dumpty, 'When I use a word, it means just what I choose it to mean – neither more nor less' (Carroll, 1872).

Although strongly decaying particles cannot be observed directly, they do produce readily observable effects. For instance, consider a particle $A$ decaying by

$$A \rightarrow B + C \tag{40.1}$$

The particle $A$ will show up as a resonance in the scattering of $B$ by $C$

$$B + C \rightarrow A \rightarrow B + C \tag{40.2}$$

as illustrated in Fig. 40.1. Similar resonances occur in atomic physics and in nuclear physics (see, for instance, Eisberg, 1961). According to the Heisenberg uncertainty principle

$$\Delta E \, \Delta t \sim \hbar \tag{40.3}$$

where $\Delta t$ is the lifetime and $\Delta E$ the width of the resonance. $\Delta E$ is the width at half the height, as shown in Fig. 40.1.

A scattering resonance is analogous to a resonance of a classical oscillator (see Appendix E). The variation of the resonant scattering cross-section with energy has the same form as the variation of response of an oscillator with frequency. The resonant scattering cross-section is given by the Breit–Wigner formula

$$\sigma_{\text{resonance}} = \frac{\text{constant}}{(E - E_0)^2 + \Gamma^2/4} \tag{40.4}$$

where $E_0 = M_A c^2$, and $\Gamma$ is the width, i.e. $\Gamma = \Delta E$ (Halliday, 1950).

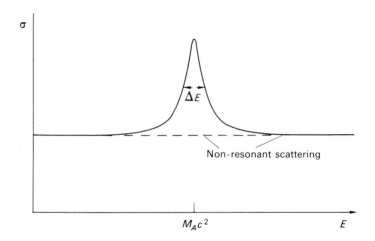

FIGURE 40.1    Resonance for scattering $B + C \rightarrow A \rightarrow B + C$. $E$ is the total energy in the centre-of-mass frame.

The situation described by equation (40.4) and Fig. 40.1 is grossly oversimplified. The observation of a scattering resonance is usually complicated by the presence of other resonances nearby and by interference with the non-resonant scattering. Resonances in pion–nucleon scattering are discussed in Section 41.

The existence of strongly decaying particles or resonances can also be deduced by analysing the energy–momentum correlations of the decay products of a reaction. For example, consider the reaction

$$\bar{p} + p \rightarrow \pi^+ + \pi^+ + \pi^- + \pi^- + \pi^0$$

From energy–momentum correlations it can be inferred that this reaction sometimes takes place through the formation of an intermediate particle, called $\omega^0$, i.e.

$$p + \bar{p} \rightarrow \omega^0 + \pi^+ + \pi^-$$

followed by

$$\omega^0 \rightarrow \pi^+ + \pi^- + \pi^0$$

This method of observing resonances will be discussed further in Section 42.

## 41  Resonances in pion–nucleon scattering

Resonances are apparent in the total cross-sections of pions on protons, as can be seen in Figs. 41.1, 41.2 and 41.3. The total cross-section is obtained by measuring the attenuation of a pion beam which passes through a hydrogen target. The symbol $\Delta$ is used for resonances with isospin $I = \frac{3}{2}$; and for resonances with the same isospin as the nucleon, $I = \frac{1}{2}$, the same symbol, $N$, is used as for the nucleon. In some books and papers, resonances with $I = \frac{3}{2}$ and $I = \frac{1}{2}$ are designated by $N_{3/2}$ and $N_{\frac{1}{2}}$ respectively.

The resonances are usually designated by the appropriate symbol, $\Delta$ or $N$, followed by the mass in MeV, e.g. $\Delta(1236)$, $N(1520)$.

The system of $\pi^+ + p$ has $I_3 = \frac{3}{2}$ and so has $I = \frac{3}{2}$. The system of $\pi^- + p$ has $I_3 = \frac{1}{2}$, and so is partly a state with $I = \frac{3}{2}$ and partly a state with $I = \frac{1}{2}$. So that resonances occurring for $\pi^+ p$ have $I = \frac{3}{2}$, and occur also for $\pi^- p$. Resonances occurring for $\pi^- p$ and not for $\pi^+ p$ have $I = \frac{1}{2}$. The resonances can be seen more clearly by examining the total cross-sections in pure isospin states. The $I = \frac{3}{2}$ cross-section is, of course, the same as the $\pi^+ p$ total cross-section. The $I = \frac{1}{2}$ cross-section is given by

$$\sigma_{\frac{1}{2}} = \tfrac{3}{2}\sigma_{\text{tot}}^- - \tfrac{1}{2}\sigma_{\text{tot}}^+ \tag{41.1}$$

where $\sigma_{\text{tot}}^-$, $\sigma_{\text{tot}}^+$ are the total cross-sections for $\pi^- p$ and $\pi^+ p$ respectively. (See Exercise 4.) The pion–nucleon total cross-sections in the two isospin states are shown in Fig. 41.4. The $I = \frac{1}{2}$ total cross-section is shown in more detail for higher energies in Fig. 41.5.

The total cross-sections can be analysed in terms of Breit–Wigner shapes and non-resonant background, and the widths of the resonances can be obtained in this way. However, it must be

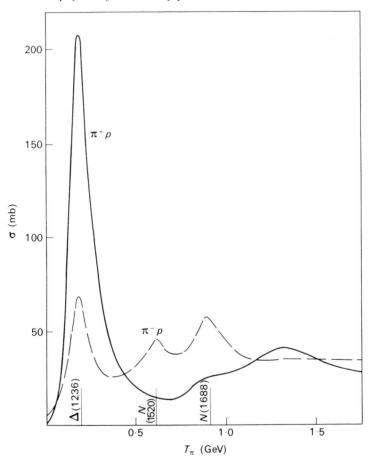

admitted that the resonances in Figs. 41.2, 41.4 and 41.5 do not seem to the eye to bear too great a resemblance to the Breit–Wigner shape shown in Fig. 40.1.

The pion–nucleon resonances can be studied more effectively by examining, as well as the total cross-section, the other experimental data on the pion–nucleon interaction such as the angular distribution of scattered pions and the polarization of the recoiling proton. Many resonances that are not apparent in the total cross-section have been discovered, and the spins and parities of many resonances have been determined from the detailed investigation of collisions of pions with protons. Figure 41.6 shows some experimental results on the angular

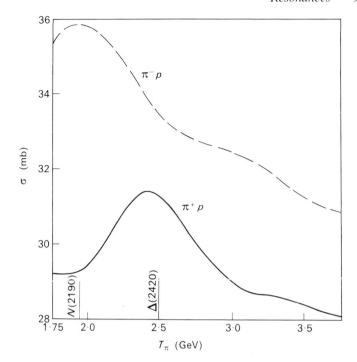

FIGURE 41.1   Total cross-sections for the scattering of pions by protons. $T_\pi$ is the kinetic energy of the pion in the laboratory frame. The positions of some prominent resonances are marked, and their masses given in MeV. Resonances labelled by $N$ have isospin $I=\frac{1}{2}$, and resonances labelled by $\Delta$ have $I=\frac{3}{2}$. The curves are from a compilation of data by Barashenkov (1968).

distribution of pions elastically scattered by protons and of the charge exchange reaction

$$\pi^- + p \rightarrow \pi^0 + n$$

The results on the pion–nucleon interaction can be summarized in terms of the pion–nucleon cross-section in states of definite isospin $I$, orbital angular momentum $l$ and total spin $J$ (the sum of the orbital angular momentum $l$ and the spin of the nucleon, $J = l \pm \frac{1}{2}$). These states are usually specified by the symbol

$$l_{2I,\,2J} \quad \text{or} \quad l2I\ 2J$$

where the spectroscopic notation is used for $l$, i.e.

$$l = 0, \quad 1, \quad 2, \quad 3, \quad 4, \quad \ldots$$
$$\quad\; S \quad P \quad D \quad F \quad G \quad \text{etc.}$$

For instance, the state $D35$ has $l=2$, $I=\frac{3}{2}$, $J=\frac{5}{2}=l+\frac{1}{2}$. The state of a plane wave of pions incident on protons can be expanded in terms

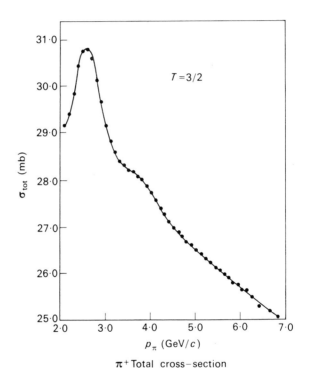

$\pi^+$ Total cross-section

FIGURE 41.2 $\pi^+p$ total cross-section. Resonances $\Delta(2420)$ and $\Delta(2850)$ are apparent at $p_\pi = 2{\cdot}65$ GeV/$c$ and $3{\cdot}84$ GeV/$c$ respectively. (From Citron *et al.*, 1966.) $p_\pi$ is the momentum of the incident pion in the laboratory frame. (See exercise 2.)

of the states $l\ 2I\ 2J$, which are called partial waves. The cross-sections for the states $l\ 2I\ 2J$ are called partial wave cross-sections. Figure 41.7 shows the results of an analysis by Bareyre, Bricman and Villet (1968) for the partial wave cross-sections for the pion-nucleon interaction. The partial wave in which a resonance occurs determines the spin and parity of the resonance. The spin is $J$ and the parity is $(-1)^l$. The resonances stand out more clearly from one another in the curves for partial wave cross-sections, Fig. 41.7, than in the curves for total cross-sections, Fig. 41.4.

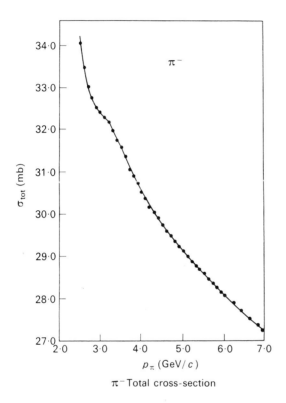

$\pi^-$ Total cross-section

FIGURE 41.3   $\pi^- p$ total cross-section. (From Citron *et al.*, 1966.)

A detailed and readable account of the pion–nucleon interaction is given by Cence (1969). Some of the pion–nucleon resonances are listed in Table 41.1. A full critical list is given in *Review of Particle Properties*, which is a review of the properties of leptons, mesons and baryons by the Particle Data Group which is updated and published about once a year. Two of the reviews are Rittenberg *et al.* (1971) and Söding *et al.* (1972). However, the reader should consult the latest *Review of Particle Properties*.

TABLE 41.1    Pion–nucleon resonances. Abridged from Söding *et al.* (1972)

|  | $I$ | $J^P$ |  | Mass (MeV) | Width, $\Gamma$ (MeV) |
|---|---|---|---|---|---|
| $N(1470)$ | $\frac{1}{2}$ | $\frac{1}{2}^+$ | $P11$ | 1435–1505 | 165–400 |
| $N(1520)$ | $\frac{1}{2}$ | $\frac{3}{2}^-$ | $D13$ | 1510–1540 | 105–150 |
| $N(1535)$ | $\frac{1}{2}$ | $\frac{1}{2}^-$ | $S11$ | 1500–1600 | 50–160 |
| $N(1670)$ | $\frac{1}{2}$ | $\frac{5}{2}^-$ | $D15$ | 1655–1680 | 105–175 |
| $N(1688)$ | $\frac{1}{2}$ | $\frac{5}{2}^+$ | $F15$ | 1680–1692 | 105–180 |
| $N(1700)$ | $\frac{1}{2}$ | $\frac{1}{2}^-$ | $S11$ | 1665–1765 | 100–400 |
| $N(2190)$ | $\frac{1}{2}$ | $\frac{7}{2}^-$ | $G17$ | 2000–2260 | 270–325 |
| $N(2650)$ | $\frac{1}{2}$ | $?^-$ |  | 2650 | 360 |
| $N(3030)$ | $\frac{1}{2}$ | $?$ |  | 3030 | 400 |
| $\Delta(1236)$ | $\frac{3}{2}$ | $\frac{3}{2}^+$ | $P33$ | 1230–1236 | 110–122 |
| $\Delta(1650)$ | $\frac{3}{2}$ | $\frac{1}{2}^-$ | $S31$ | 1615–1695 | 130–200 |
| $\Delta(1950)$ | $\frac{3}{2}$ | $\frac{7}{2}^+$ | $F37$ | 1930–1980 | 140–220 |
| $\Delta(2420)$ | $\frac{3}{2}$ | $\frac{11}{2}^+$ |  | 2320–2450 | 270–350 |
| $\Delta(2850)$ | $\frac{3}{2}$ | $?^+$ |  | 2850 | 400 |
| $\Delta(3230)$ | $\frac{3}{2}$ | $?$ |  | 3230 | 440 |

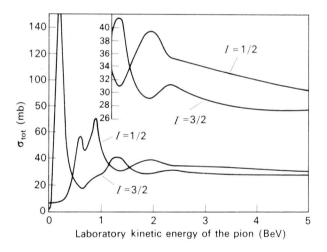

FIGURE 41.4    The pion–nucleon total cross-section in the two isospin states. For pion kinetic energies greater than 1·2 BeV, the cross-section is also shown magnified 10 times. The following resonances are apparent at the listed values of $T_\pi$, the pion kinetic energy.

$N(1520)\ T_\pi = 0\cdot61$ BeV;    $N(1688)\ T_\pi = 0\cdot90$ BeV;
$N(2190)\ T_\pi = 1\cdot94$ BeV;    $\Delta(1236)\ T_\pi = 0\cdot20$ BeV;
$\Delta(1950)\ T_\pi = 1\cdot41$ BeV;    $\Delta(2420)\ T_\pi = 2\cdot4$ BeV

(from Diddens *et al.*, 1963).

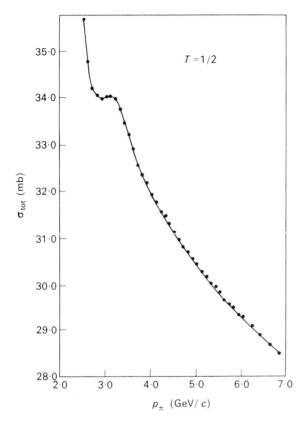

FIGURE 41.5 The total cross-section in the isospin-$\frac{1}{2}$ state. The resonance $N(2650)$ is apparent at $p_\pi = 3\cdot26$ GeV/$c$. $p_\pi$ is the momentum of the incident pion in the laboratory frame. (From Citron *et al.*, 1966.) (*Note*: the figure uses $T$ for isospin.)

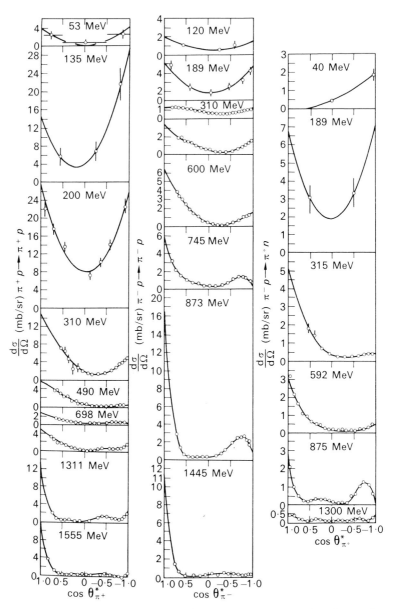

FIGURE 41.6   Differential cross-sections at selected energies for $\pi^\pm p \to \pi^\pm p$ and $\pi^- p \to \pi^0 n$. (From Cence, 1969.)

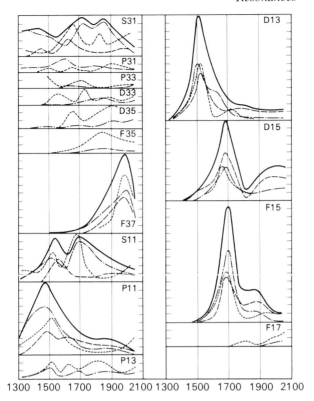

1300  1500  1700  1900  2100 1300  1500  1700  1900  2100

FIGURE 41.7   Partial wave cross-sections. $\sigma_{tot}$, solid curves; $\sigma_{el}$, dot-dashed curves; $\sigma_{inel}$, dashed curves. The abscissa is the total centre-of-mass energy in MeV. This figure should be compared with Fig. 41.4. Several additional resonances are apparent here. (From Bareyre *et al.*, 1968.)

## 42  Detection of resonance particles by energy–momentum correlations

The class of resonances that can be detected in scattering experiments is very restricted. For instance, resonances that might occur for the scattering of pions by $\Lambda$ particles could not be observed directly in $\pi\Lambda$ scattering as we do not have any target of $\Lambda$ particles. However, such resonances can be detected by other methods using the correlations between the energy and momenta of particles emitted after a reaction.

Consider a particle $A$ with two possible modes of decay, one a two-particle decay

$$A \rightarrow B + C \qquad (42.1)$$

and the other a three-particle decay

$$A \rightarrow B + D + E \qquad (42.2)$$

We consider the case where the only particle observed in the final state is particle $B$.

In the two-particle decay, the energy of particle $B$ is completely determined by conservation of energy and momentum. For instance, in the rest frame of $A$, the kinetic energy of particle $B$ is given by

$$T_B = \frac{\{(M_A - M_B)^2 - M_C^2\}c^2}{2M_A} \qquad (42.3)$$

(see Chapter 6, Exercise 2).

On the other hand, in the three-body decay, the laws of conservation and momentum do not uniquely determine the energy of $B$, and if many events are observed a range of energies of $B$ will be seen, from $T_B = 0$ to

$$T_B = \frac{\{(M_A - M_B)^2 - (M_D + M_E)^2\}c^2}{2M_A} \qquad (42.4)$$

The distribution of $T_B$ for many decays can be estimated from statistical theory. The probability of the particle $B$ being emitted with a particular energy $T_B$ depends on the amount of phase space available to the combination of particles $B$, $D$ and $E$ in which the kinetic energy of $B$ is $T_B$; or to put it another way, depends on the number of final states in which $B$ has that particular kinetic energy. (In the literature, such estimates from statistical theory for the number of particles with a particular energy are frequently labelled 'phase space', e.g. see Fig. 45.1.) In this way, one obtains the distribution of $T_B$ of the kind shown in Fig. 42.1b.

By observing the distribution of $T_B$ for many decays of $A$ it is possible to determine whether $A$ decays to three particles or to two particles according to whether the pattern of Fig. 42.1b or Fig. 42.1a is observed.

We now consider the case when

$$M_C > M_D + M_E \qquad (42.5)$$

and the following decay can occur

$$C \rightarrow D + E \qquad (42.6)$$

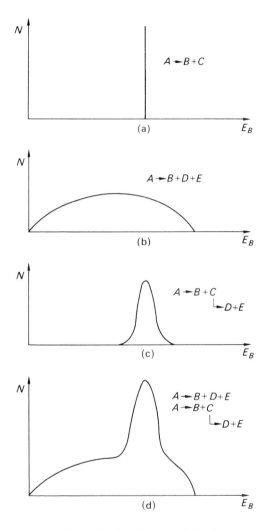

FIGURE 42.1   The distribution of $T_B$, the energy of particle $B$, in the decay of particle $A$ for several possible decay modes for $A$. The observation of a distribution of the type shown in figures (c) and (d) demonstrates the existence of a resonance particle $C$ and enables the determination of its mass $M_C$.

We consider the case when $C$ is a resonance particle and decays so rapidly that the presence of $C$ cannot be directly observed. Then the decay (42.2) may proceed directly or in a two-step process

$$A \rightarrow B + C$$
$$\phantom{A \rightarrow B + } \llcorner\rightarrow D + E \qquad (42.7)$$

$C$ does not live long enough to leave a track in a bubble chamber. However, the existence of $C$ can be inferred by observing many decays of $A$.

If the decay of $A$ occurs as in equation (42.7), the distribution of $T_B$ will not be as in Fig. 42.1b. If $C$ had an infinite lifetime, the energy of $E_B$ would be uniquely determined as the decay of $A$ would be a two-body decay. As $C$ has a finite lifetime, and so by the uncertainty principle its rest energy or mass has a finite width, then the distribution of $T_B$ would be as in Fig. 42.1c. In general, the decay of $A$ would proceed both directly as given by equation (42.2) and as a two-stage process given by equation (42.7), and so the distribution of $T_B$ will be as shown in Fig. 42.1d. In this case, measurement of $T_B$ for many decays still enables us to establish the existence of the resonance particle $C$ and to determine its mass $M_C$, since corresponding to $B$ particles having energy $T_B$ we obtain from equation (42.3)

$$M_C = [(M_A - M_B)^2 - 2M_A T_B c^{-2}]^{\frac{1}{2}} \qquad (42.8)$$

In practice, the usual experiment detecting a resonance is slightly more complicated than the case just dealt with, in that the initial state is not that of a single particle $A$ but a state of two particles $F$ and $G$, i.e. the experiment involves a reaction

$$F + G \rightarrow B + D + E \qquad (42.9)$$

$$F + G \rightarrow B + C$$
$$\phantom{F + G \rightarrow B + } \llcorner\rightarrow D + E \qquad (42.10)$$

The above treatment still holds when we replace the particle $A$ by a state of $F$ and $G$ with some definite energy, and by examining the events occurring when a beam of $F$ particles is incident on a target of $G$ particles it can be seen whether a resonance particle such as $C$ occurs.

An example of this procedure is the discovery of the $Y_1^*$ resonance by Alston *et al.* (1960) in the reaction

$$K^- + p \rightarrow \Lambda^0 + \pi^+ + \pi^- \qquad (42.11)$$

produced by 1·15 GeV/$c$ $K^-$-mesons. The experimental results are shown in Fig. 42.2. The two histograms in the diagram are diagrams of the same type as those in Fig. 42.1, being plots of number of events versus the kinetic energy of one particle. The solid curves show the predictions of statistical theory and it is seen that the experimental results deviate considerably from these curves indicating that there is a resonance particle occurring in the reaction. The reaction occurs in two stages

$$K^- + p \rightarrow \pi^+ + Y_1^{*-}$$
$$\quad\quad\quad \lfloor_{\,\rightarrow\,\, \pi^- + \Lambda^0} \quad\quad\quad (42.12)$$

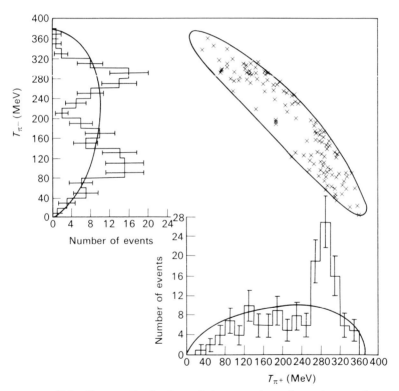

FIGURE 42.2 Energy distribution of the two pions from the reaction $K^- + p \rightarrow \Lambda + \pi^+ + \pi^-$. Each event is plotted only once on the Dalitz plot, which should be uniformly populated if phase space dominated the reaction. The two energy histograms are merely one-dimensional projections of the two-dimensional plot, and each event is represented once on each histogram. The solid lines superimposed over the histograms are the phase–space curves. (From Alston *et al.*, 1960.)

or

$$K^- + p \rightarrow \pi^- + Y_1^{*+}$$
$$\phantom{K^- + p \rightarrow \pi^-}\big\downarrow \pi^+ + \Lambda^0 \qquad (42.13)$$

Figure 42.2 also shows the Dalitz plot for the reaction. In a Dalitz plot, each event is shown as a point on a two-dimensional plot, the axes corresponding to appropriate energies observed in the reaction, in this case as shown in Fig. 42.2 to the kinetic energies of the emitted pions as measured in the centre-of-mass frame. Statistical theory predicts that, for a final state of three particles, the events in a Dalitz plot should be uniformly distributed in the kinematically allowed region, the boundary of which is shown in Fig. 42.2.

In Fig. 42.2, the histograms for both $\pi^-$ and $\pi^+$ show peaks at a kinetic energy of $T_\pi = 280$ MeV corresponding to a resonance particle with a mass of 1385 MeV. The $\pi^+\Lambda$ and the $\pi^-\Lambda$ resonances at the same energy are two different charge states of a particle now called the $\Sigma(1385)$; in older papers it is called the $Y_1^*(1385)$. Since it occurs as a resonance in a strong reaction conserving strangeness, the $\pi\Lambda$ resonance has strangeness $S = -1$. As this resonance has charge states $Q = +1$ and $Q = -1$, the value of its isospin is at least $I = 1$. Subsequent experiments have shown the 1385 MeV $\pi\Lambda$ resonance has $I = 1$. Resonances with $S = -1$ were designated by $Y^*$ with a suffix denoting the value of $I$. It is now more usual to denote $S = -1$, $I = 1$ resonances by $\Sigma$ and $S = -1$, $I = 0$ resonances by $\Lambda$.

There is a second peak in the histogram for $\pi^-$ at a $\pi^-$ kinetic energy of $T_{\pi^-} = 100$ MeV. Inspection of the Dalitz plot shows that this second peak is mainly due to events associated with the $\pi^-\Lambda$ resonance, and thus this second peak is not necessarily evidence for another $\pi\Lambda$ resonance. The presence of resonances is seen much more clearly on a Dalitz plot than on a histogram for the energy of only one particle.

Instead of plotting data in terms of the kinetic energy $T_B$ of one particle, the invariant mass $M_{DE}$ of the combination of the two other particles $D$ and $E$ may be used. The two variables are related by equation (42.8), replacing $M_C$ by $M_{DE}$. $M_{DE}$ is also given by

$$M_{DE} = c^{-2}[E_{DE}^2 - c^2 \mathbf{p}_{DE}^2]^{\frac{1}{2}} \qquad (42.14)$$

where

$$E_{DE} = E_D + E_E \qquad (42.15)$$

$$\mathbf{p}_{DE} = \mathbf{p}_D + \mathbf{p}_E \qquad (42.16)$$

Since $M_{DE}$ is a relativistic invariant, it can be obtained from equations (42.14), (42.15) and (42.16) in any inertial frame of reference. $M_{DE}$ is also called the effective mass of the combination $DE$.

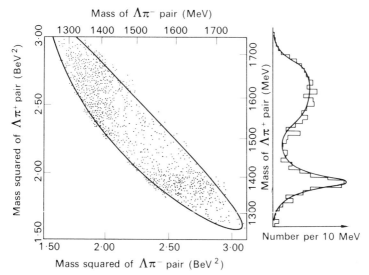

FIGURE 42.3   Dalitz plot of $\Lambda\pi^+\pi^-$ events from $K^-p$ interactions at 1·22 BeV/$c$. The square of $\Lambda\pi^+$ effective mass is plotted against the square of $\Lambda\pi^-$ effective mass. Scales giving the masses in MeV are also shown. Projection of the events onto the $\Lambda\pi^+$ mass axis is displayed to the right of the figure; the curve represents the fitting of Breit–Wigner resonance expressions to the $\Lambda\pi^+$ and $\Lambda\pi^-$ systems. (From Shafer *et al.*, 1963.)

If events for a three-particle final state are plotted in a Dalitz plot according to the squares of the invariant masses of two of the possible two-particle combinations, statistical theory would predict that the density of events should be uniform in the kinematically allowed region. Figure 42.3 shows the data of Shafer *et al.* (1963) on the 1385 MeV $\pi\Lambda$ resonance observed in the reaction $K^- + p \rightarrow \Lambda + \pi^+ + \pi^-$. Figure 42.3 should be compared with Fig. 42.2.

## 43 More baryon resonances

A resonance at 1530 MeV with $S = -2$ and $I = \frac{1}{2}$ was seen by Pjerrou *et al.* (1962) and by Bertanza *et al.* (1962). In Figs. 43.1 and 43.2, the experimental results of Schlein *et al.* (1963) are shown for the reaction

$$K^- + p \rightarrow \Xi^- + \pi^+ + K^0$$

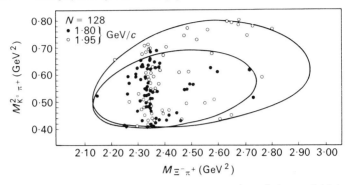

FIGURE 43.1    Dalitz plot for $K^- + p \to \Xi^- + \pi^+ + K^0$. (From Schlein *et al.*, 1963.)

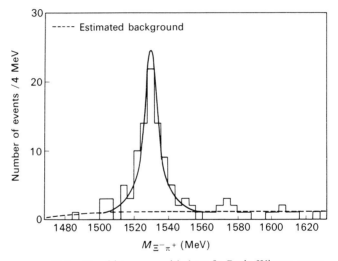

FIGURE 43.2    $M_{\Xi\pi}$ histogram with best-fit Breit–Wigner curve folded with experimental resolution curve, for the reaction $K^- + p \to \Xi^- + \pi^+ + K^0$. (From Schlein *et al.*, 1963.)

for $K^-$-mesons of 1·80 GeV/$c$ and 1·95 GeV/$c$ demonstrating the existence of a $\pi^+ \Xi^-$ resonance at 1530 MeV. As this resonance has the same strangeness and isospin as the $\Xi$ it is called the $\Xi(1530)$.

Most hyperon resonances were observed in Dalitz plots of three-body final states. But Fig. 43.3 shows how a resonance was observed as a peak in the cross-sections for the reactions

$$K^- + p \to \Lambda + \pi^+ + \pi^-$$
$$K^- + p \to \overline{K^0} + n$$

(Ferro–Luzzi, 1962.) This $K^-p$ resonance occurs at 1520 MeV and has $S = -1, I = 0$, and is called the $\Lambda(1520)$. It has also been called the $Y_0^*(1520)$. The $\Lambda(1520)$ has also been seen in Dalitz plots of three-body final states.

There are many more baryon resonances and more are found each year. Some of these resonances are listed in Appendix G. For a complete list of resonances, see the latest Particle Data Group *Review of Particle Properties* (see end of Section 41).

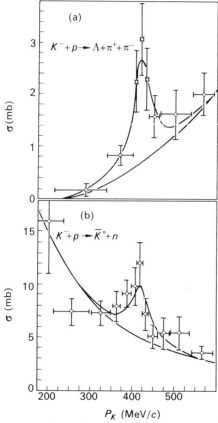

FIGURE 43.3 Momentum dependence of the cross-section for the reactions (a) $K^- + p \rightarrow \Lambda + \pi^+ + \pi^-$ and (b) $K^- + p \rightarrow \bar{K}^0 + n$ showing resonance corresponding to $\Lambda(1520)$. The lower curves in (a) and (b) represent the presumed nonresonant backgrounds, while the upper curves contain in addition the superposed resonance. (From Ferro-Luzzi *et al.*, 1962.)

## 44 The discovery of the $\Omega^-$

Among the baryon resonances, those with $J^P = \frac{3}{2}^+$ are of special interest. They are

$$\Delta(1236)\ S = 0,\ I = \tfrac{3}{2} \qquad \text{with 4 charge states}$$
$$\Sigma(1385)\ S = -1,\ I = 1 \quad \text{with 3 charge states}$$
$$\Xi(1530)\ S = -2,\ I = \tfrac{1}{2} \quad \text{with 2 charge states}$$

$M_\Sigma - M_\Delta = 149$ MeV and $M_\Xi - M_\Sigma = 145$ MeV are approximately the same. The $J^P = \frac{3}{2}^+$ baryons are arrayed in Fig. 44.1 and the array is completed by a particle called the $\Omega^-$ hyperon. The array of Fig. 44.1 has a group-theoretical significance, as will be discussed more fully in Section 51, and the masses of the different members of the array are expected to be equally spaced. From this pattern, the existence of the $\Omega^-$ hyperon was predicted with a mass of about 1680 MeV and quantum numbers $J^P = \frac{3}{2}^+$, $I = 0$, $S = -3$. Such a particle cannot decay by the strong or electromagnetic interactions into any known particles, as, for any combination of known particles with total strangeness $S = -3$, the total of the rest masses is greater than the rest mass of the $\Omega^-$. For instance

$$\Omega^- \nrightarrow \Xi^- + K^0$$

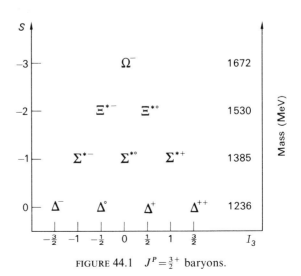

FIGURE 44.1   $J^P = \frac{3}{2}^+$ baryons.

since

$$M_{\Xi^-} + M_{K^0} = 1809 \text{ MeV}$$

considerably greater than the mass of the $\Omega^-$. The decay of the $\Omega^-$ cannot conserve strangeness and so must proceed by the weak interactions. Thus the $\Omega^-$ has a long enough lifetime to be observed as a track in a bubble chamber.

Following the theoretical prediction of the existence of the $\Omega^-$, the $\Omega^-$ was discovered by Barnes *et al.* (1964) at the Brookhaven National Laboratory. An interesting account of this experiment is given by Fowler and Samios (1964). Figure 44.2 shows the photograph and line diagram of the event observed in the hydrogen bubble chamber exposed to a beam of 5·0 BeV/$c$ $K^-$-mesons. The observed event was interpreted as

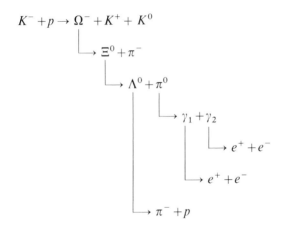

Other decay modes of the $\Omega^-$ have since been observed,

$$\Omega^- \rightarrow \Xi^- + \pi^0$$
$$\Omega^- \rightarrow \Lambda^0 + K^-$$

and the mass of the $\Omega^-$ has been measured as

$$M_{\Omega^-} = (1672 \cdot 5 \pm 0 \cdot 5) \text{ MeV}$$

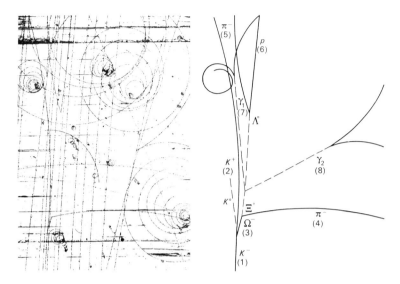

FIGURE 44.2    Photograph and line diagram of event showing decay of $\Omega^-$. (From Barnes *et al.*, 1964.)

## 45  Meson resonances with $S=0$

Figure 45.1 shows the results of Erwin *et al.* (1961) for the reactions

$$\pi^- + p \rightarrow \pi^+ + \pi^- + n \qquad (45.1)$$

$$\rightarrow \pi^- + \pi^0 + p \qquad (45.2)$$

demonstrating the existence of a $\pi$–$\pi$ resonance called the $\rho$-meson. The $\rho$-meson has $J^P = 1^-$, $I = 1$, $S = 0$, and $M_\rho = 765$ MeV. Figure 45.1 shows the combined mass spectrum for the $\pi^- \pi^0$ and $\pi^- \pi^+$ system. Peaks appeared in both the $\pi^- \pi^0$ and $\pi^- \pi^+$ mass spectra, but the data were combined for better statistics. The effective mass of the two pions is given by equations (42.14), (42.15) and (42.16). The peak in the mass spectrum for the two pions occurs at 765 MeV. Thus the reactions given by equations (45.1) and (45.2) occur at least partly as

$$\pi^- + p \rightarrow \rho^0 + n$$
$$\qquad\qquad \rightarrow \pi^+ + \pi^- \qquad (45.3)$$

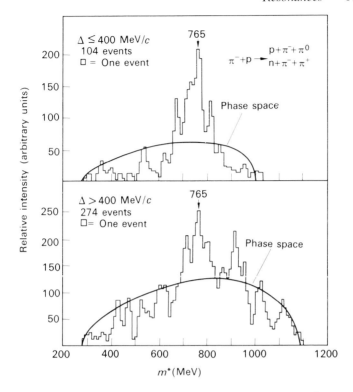

FIGURE 45.1 The combined mass spectrum for the $\pi^- \pi^0$ and $\pi^- \pi^+$ system showing the $\rho$ meson at 765 MeV. The events are divided into two cases, those with $\Delta \leqslant 400$ MeV/$c$ and those with $\Delta > 400$ MeV/$c$ where $\Delta$ is the four-momentum transfer to the nucleon. The smooth curve labelled 'phase space' is the prediction of statistical theory as modified for the included momentum transfer and normalized to the number of events plotted. (From Erwin *et al.*, 1961.)

$$\pi^- + p \rightarrow \rho^- + n$$
$$\phantom{\pi^- + p \rightarrow} \big\downarrow \rightarrow \pi^- + \pi^0 \qquad (45.4)$$

In subsequent experiments, the $\rho^+$ has also been observed.

Another mesonic resonance is the $\omega$-meson with $S=0$, $I=0$ and $J^P = 1^-$. Maglić *et al.* (1961) searched for the $\omega$-meson as a three-pion resonance among the pions produced from the annihilation of antiprotons with protons when a beam of antiprotons of momentum

1·61 BeV/$c$ was incident on a hydrogen bubble chamber. The search was made assuming $M_\omega > 3M_\pi$, so that the decay mode

$$\omega \rightarrow \pi^+ + \pi^- + \pi^0 \qquad (45.5)$$

is possible. Such a three-pion decay mode was searched for by studying the mass distribution of triplets of pions in the reaction

$$\bar{p} + p \rightarrow \pi^+ + \pi^+ + \pi^- + \pi^- + \pi^0 \qquad (45.6)$$

For 2500 four-prong events, the energy and momentum of each charged pion was measured, and the presence of the $\pi^0$ was deduced for 800 events using the laws of conservation of energy and momentum. For each of the 800 events, the three-body effective mass

$$M_3 = [(E_1 + E_2 + E_3)^2 + (\mathbf{p}_1 + \mathbf{p}_2 + \mathbf{p}_3)^2 c^2]^{\frac{1}{2}} c^{-2} \qquad (45.7)$$

was calculated for each pion triplet in the reaction (45.6). Of course, even if a resonance does occur, there is no way of knowing beforehand which three of the five pions make up the resonance particle, so that each possible combination of three pions had to be considered. For each event, there are 10 combinations of three pions, corresponding to the following charge states

$$|Q| = 1 : \pi^\pm \pi^\pm \pi^\mp \text{ (4 combinations) } \quad A$$
$$|Q| = 2 : \pi^\pm \pi^\pm \pi^0 \text{ (2 combinations) } \quad B$$
$$|Q| = 0 : \pi^+ \pi^- \pi^0 \text{ (4 combinations) } \quad C$$

Figure 45.2 shows the $M_3$ distribution for the 800 events. The solid curves are an approximation to statistical theory. The distributions $A$ and $B$ have no resonance peaks, indicating that there is no correlation between these particular triplets of $\pi$-mesons. Distribution $C$ shows a peak at 787 MeV, demonstrating the existence of a neutral $\omega$-meson. The large background to the peak is due to the fact that only one quarter of the triplets plotted in $C$ can possibly be associated with the $\omega^0$ and that the reaction does not always proceed through the formation of a resonance particle. That the peak corresponding to the $\omega$ occurs only for distribution $C$ corresponding to charge state $Q=0$ and not for distribution $A$, $|Q|=1$, or for $B$, $|Q|=2$, is consistent with the assignment of isospin $I=0$ to the $\omega$ as an isospin singlet. Reaction (45.6) proceeds at least partly by

$$\bar{p} + p \rightarrow \omega^0 + \pi^+ + \pi^-$$
$$\phantom{\bar{p} + p \rightarrow \omega^0}\big\downarrow \pi^+ + \pi^- + \pi^0 \qquad\qquad (45.8)$$

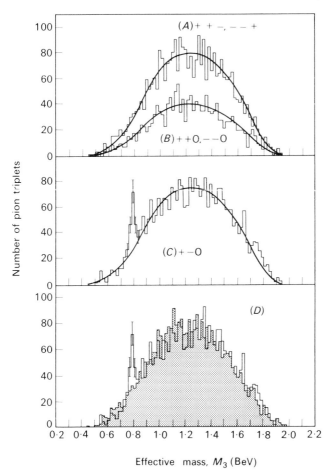

FIGURE 45.2 Number of pion triplets versus effective mass $M_3$ of the triplets for reaction

$$\bar{p} + p \rightarrow 2\pi^+ + 2\pi^- + \pi^0$$

$A$ is the distribution for the combination $|Q| = 1$; $B$ is for the combination $|Q| = 2$; and $C$ is for $Q = 0$; with 3200, 1600 and 3200 triplets respectively. In $D$, the combined distributions $A$ and $B$ (shaded area) are contrasted with distribution $C$ (heavy line). (From Maglić *et al.*, 1961.)

The $\omega$-meson showed more clearly in the experiment of Alff *et al.* (1962, 1966) on the production of pion resonances in $\pi^+ p$ reactions. In this experiment, the $\rho$-meson and the $\eta$-meson were also produced. The $\eta$-meson has $J^P = 1^-$, $S = 0$, $I = 0$ and mass $M_\eta = 549$ MeV. Figure 45.3a shows the mass spectrum of the $\pi^+ \pi^0$ combinations in the reaction

$$\pi^+ + p \rightarrow \pi^+ + \pi^0 + p \qquad (45.9)$$

and Fig. 45.3b shows the mass spectrum of the $\pi^+ \pi^-$ combinations in the reaction

$$\pi^+ + p \rightarrow \pi^+ + \pi^- + \pi^+ + p \qquad (45.10)$$

Breit–Wigner resonance curves have been superimposed on the phase-space backgrounds for the two-pion distribution. The peaks show that some of the events proceeded by the production of the $\rho$-meson,

$$\pi^+ + p \rightarrow \rho^+ + p$$
$$\qquad \quad \Big\downarrow \rightarrow \pi^+ + \pi^0 \qquad (45.11)$$

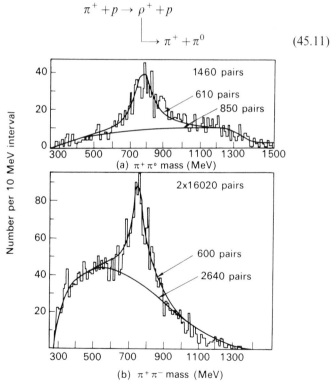

FIGURE 45.3    Mass distributions (a) for $\pi^+ \pi^0$ pairs from $\pi^+ + p \rightarrow \pi^+ + p + \pi^0$ events, (b) for $\pi^+ \pi^-$ pairs from $\pi^+ + p \rightarrow \pi^+ + p + \pi^+ + \pi^-$ events and (c) for $\pi^+ \pi^- \pi^0$ triplets from $\pi^+ + p \rightarrow \pi^+ + p + \pi^+ + \pi^- + \pi^0$ events.

$$\pi^+ + p \rightarrow \rho^0 + \pi^+ + p$$
$$\phantom{\pi^+ + p \rightarrow \rho^0 +} \mathrel{\raise0pt\hbox{$\llcorner$}}\!\!\rightarrow \pi^+ + \pi^-. \qquad (45.12)$$

Figure 45.3c shows the distribution for the effective mass of triplets $\pi^+ \pi^- \pi^0$ from

$$\pi^+ + p \rightarrow \pi^+ + \pi^+ + \pi^- + \pi^0 + p \qquad (45.13)$$

events. A phase-space curve has been drawn for events outside the peaks on the triple-pion distribution. The peak with 800 triplets at 782 MeV corresponds to the $\omega^0$-meson, and the peak with 100 triplets at 548 MeV corresponds to the $\eta^0$-meson.

Evidence for many meson resonances with $S = 0$ has been obtained at CERN with an experimental arrangement suggested by Maglić and Gosta (1965) and called the missing mass spectrometer. Consider the reaction

$$\overset{1}{\pi^-} + \overset{2}{p} \rightarrow \overset{3}{p} + N\pi \qquad (45.14)$$

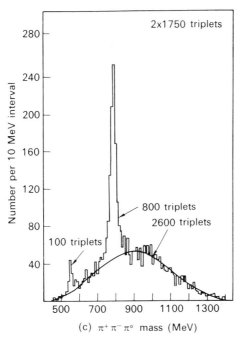

(c) $\pi^+ \pi^- \pi^0$  mass (MeV)

The peak in (a) and (b) corresponds to the $\rho$. In (c) the peak with 800 triplets corresponds to the $\omega^0$, and the peak with 100 triplets corresponds to the $\eta^0$. (From Alff *et al.*, 1962.)

where $N = 1, 2$, etc. In some cases the reaction will produce un-correlated pions. In other cases the reaction will proceed by producing a resonance $X$ which decays into $N$ pions; the reaction can then be written as

$$
\begin{array}{cccc}
1 & 2 & 3 & 4
\end{array}
$$

$$\pi^- + p \to p + X^- \tag{45.15}$$

With fixed momentum of the incident pion, $p_1$, by measuring $p_3$ and $\theta_3$, the momentum and angle to the incident pion at which the proton recoils, the effective mass of the $N$ pions can be determined as the 'missing mass' given by $M$,

$$
\begin{aligned}
M^2 &= c^{-4} \left[(\text{missing energy})^2 - c^2(\text{missing momentum})^2\right] \\
&= c^{-4} \left[(E_1 + M_2 c^2 - E_3)^2 - c^2(\mathbf{p}_1 - \mathbf{p}_3)^2\right] \tag{45.16} \\
&= c^{-4} \left[(E_1 + M_2 c^2 - E_3)^2 - c^2 p_1^2 - c^2 p_3^2 + 2 p_1 p_3 \cos \theta_3\right]
\end{aligned}
$$

For the production of uncorrelated pions, the distribution in $M$ is a smooth function with a broad minimum. However, if the $N$ pions are produced as a resonance $X$, a peak will occur in the distribution of $M$ at the mass of the resonance $M_X$. If

$$v_c > v_3'$$

where $v_c$ is the velocity of the centre-of-mass in the laboratory frame and $v_3'$ is the velocity of the recoil proton in the centre-of-mass system, the angle $\theta_3$ has a maximum less than $\pi/2$ and at this maximum of $\theta_3$, the transformation between the cross-section in the centre-of-mass frame and the cross-section in the laboratory frame is singular, as shown in Appendix A by equation (A.68). For a definite sharp value of $M_X$, there will be a singularity in the laboratory cross-section at

$$\sin \theta_3 = p_3' / M_3 c \beta \gamma$$

where

$$
\begin{aligned}
\beta &= v_c/c \\
\gamma &= (1 - \beta^2)^{\frac{1}{2}}
\end{aligned}
$$

(see equation (A.69)). Where $X$ is a resonance with a finite width, there will be a peak in the laboratory cross-section. This peak is called the Jacobian peak, since it occurs because of a singularity in the Jacobian for the transformation from centre-of-mass coordinates to laboratory coordinates.

Experimental results on the observation of the $\rho$-meson with the CERN missing mass spectrometer (Blieden *et al.*, 1965, 1966) are shown in Fig. 45.4. Figure 45.4b shows the laboratory angular

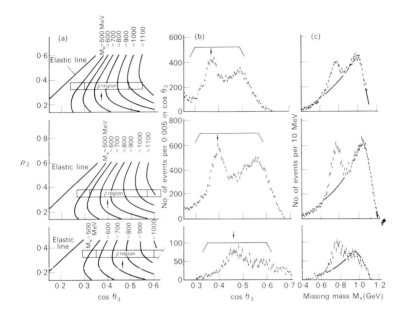

FIGURE 45.4 Observation of ρ-meson by CERN missing mass spectrometer (Blieden *et al.*, 1966). The results are for three values of the incident pion momentum, $p_1$,

    Upper row: $p_1$ = 5·0 GeV/c, 17170 events.
    Middle row: $p_1$ = 4·5 GeV/c, 29145 events.
    Lower row: $p_1$ = 3·5 GeV/c, 3069 events.

(a) Kinematic lines of the recoil proton. The horizontal heavy rectangle indicates the region in which the recoil protons are detected with full efficiency, while the light rectangle indicates the region of <100 per cent efficiency. The position for a recoil proton associated with the production of a ρ-meson,

$$\pi^- + p \rightarrow p + \rho^-$$

is indicated by arrows.
(b) Laboratory angular distributions ($N$ vs. cos $\theta_3$) of the recoil protons of all momenta within the momentum band $320 \leqslant p_3 \leqslant 380$ MeV/c. The measurement of the recoil proton momentum $p_3$ has not been used in obtaining these distributions. The $\rho^-$ enhancement (see arrows) shifts with increase of $p_1$ towards lower cos $\theta_3$, as expected from the kinematic lines in (a).
(c) Missing-mass distributions of the same data as (b), using the measured value of $p_3$ for each event, in addition to the laboratory angle $\theta_3$.

distributions of the recoil protons of all momenta within the momentum band $320 \leqslant p_3 \leqslant 380$ MeV/$c$, for three values of the incident pion momentum, $p_1$. The peak corresponding to the $\rho$-meson is indicated by an arrow, and shifts with increase of $p_1$ towards lower $\cos \theta_3$ as expected from kinematics. Figure 45.4b shows that the presence of a resonance can be deduced by observing the Jacobian peak for which only the measurement of $\theta_3$ is necessary. Using the measurement of both $\theta_3$ and $p_3$, the missing mass can be determined from equation (45.16). Figure 45.4c shows the distribution of missing mass, and the peak corresponding to the $\rho$ is clearly visible.

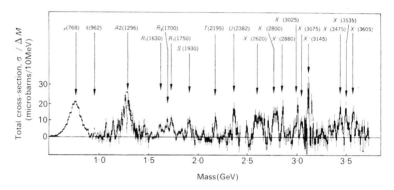

FIGURE 45.5    Non-strange boson resonances in the missing mass spectrum obtained in the reaction $\pi^- + p \rightarrow p + X^-$, where $X^-$ is the boson produced, observed by the missing-mass and boson spectrometers at CERN. The recoil proton was observed in the Jacobian peak for missing mass below 2·4 GeV and in the forward direction above this value. (From Schübelin, 1970.)

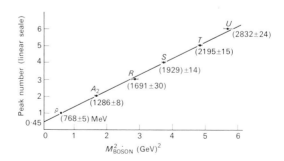

FIGURE 45.6    Number of each major peak in missing-mass spectrum in sequence of increasing mass plotted versus mass square, $M_X^2$. (From Focacci *et al.*, 1966.)

Many such measurements of the missing mass spectrum have been made, and a summary from Schübelin (1970) of the results is shown in Fig. 45.5. In an earlier review of the mass spectrum of bosons observed by the missing-mass spectrometer, Focacci *et al.* (1966) pointed out an interesting regularity between the masses of the major peaks in the missing-mass spectrum. If the major peaks from the mass spectrum are plotted in order of their mass on a linear scale versus mass square $M_X^2$, the points lie on a straight line with a slope of $1.05$ $(GeV)^2$, as shown in Fig. 45.6. The peak numbers of the $\rho$ resonance and the $A_2$ resonance in Fig. 45.6 are equal to their spins, 1 and 2, respectively. For reasons to be discussed in Chapter 12, it has been conjectured that the peak numbers of the other resonances are also equal to their spins.

## 46 Meson resonances with $S = \pm 1$

The $K^*$ with a mass of 892 MeV was the first meson resonance with $S = \pm 1$ to be established. The $K^*$ was first seen in the reactions

$$K^- + p \to K^- + \pi^0 + p \qquad (46.1)$$

$$K^- + p \to \overline{K^0} + \pi^- + p \qquad (46.2)$$

which proceed partly by

$$K^- + p \to K^{*-} + p$$
$$\qquad \qquad \big\lfloor \to K^- + \pi^0 \qquad (46.3)$$

$$K^- + p \to K^{*-} + p$$
$$\qquad \qquad \big\lfloor \to \overline{K^0} + \pi^- \qquad (46.4)$$

An example of experimental data for reaction (46.2), from Wojcicki (1964), is shown in the form of a Dalitz plot in Fig. 46.1, clearly showing the clustering of events at an effective mass for the $\overline{K^0}\pi^-$ corresponding to the mass of the $K^{*-}$. The $K^*$ has also occurred in other reactions such as

$$\pi^- + p \to \Lambda + K^{*0}$$
$$\qquad \qquad \big\lfloor \to K + \pi \qquad (46.5)$$

$$\pi^- + p \to \Sigma^0 + K^{*0}$$
$$\qquad \qquad \big\lfloor \to K + \pi \qquad (46.6)$$

$$\pi^- + p \rightarrow \Sigma^+ + K^{*+}$$
$$\llcorner \rightarrow K + \pi \qquad (46.7)$$

The $K^*$ has a spin and parity of $J^P = 1^-$.

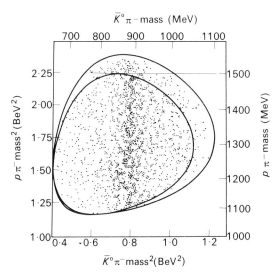

<u>FIGURE</u> 46.1    Dalitz plot for the reaction $K^- p \rightarrow \bar{K}^0 \pi^- p$ for momenta of the incident kaons $1\cdot45 < p_{K^-} < 1\cdot55$ BeV/$c$. The envelopes show the boundaries of the kinematically allowed regions for $p_{K^-} = 1\cdot45$ and $1\cdot55$ BeV/$c$. (From Wojcicki, 1964.)

Other boson resonances with $S = \pm 1$ have been observed. All such resonances are members of isospin doublets and so have $I = \frac{1}{2}$, and occur with $S = -1$ as neutral or singly positive charged, or with $S = +1$ as neutral or singly negative charged.

## 47 Resonances in various channels

Care has to be taken in the identification of a resonance particle. Not every bump in a cross-section, nor even every clustering of events in a Dalitz plot, can be interpreted as a resonance particle. In the past there have been reports of resonances which have later failed to be established as resonances. The existence of a resonance particle is more firmly established if the resonance particle is observed in several different reactions and in several different channels. Chew *et al.* (1964) have made an analogy between resonances of strongly-interacting particles and the phenomena of resonant cavities in electromagnetism. An electromagnetic cavity may have a number of wave guides attached to it. Each wave guide will allow to pass only waves of frequency greater than a particular frequency. If the electromagnetic radiation in the cavity has a frequency lower than any of the cut-off frequencies of the wave guides, then the radiation will be contained permanently in the cavity. Higher frequency radiation will have the possibility of leaking out through a number of exit wave guides. The analogous situation in particle physics can be illustrated by consideration of, for example, $\pi^+$-proton collisions. At particular values of the incident $\pi^+$ energy, the cross-section has relative maxima, corresponding to the formation of a $(\pi^+, p)$ resonance. These resonance energies are analogous to the frequencies of the electromagnetic cavity. Now, depending on how high a particular resonance energy is, there will be a certain number of modes of decay, called decay channels, available to the resonance, e.g.

$$(\pi^+, p) \to \pi^+ + p \qquad \text{at low energies}$$

$$\left.\begin{aligned} (\pi^+, p) &\to \pi^+ + p \\ &\to \rho^+ + p \\ &\to \pi^+ + \pi^0 + p \\ &\to \pi^+ + \pi^+ + n \end{aligned}\right\} \text{at high energies}$$

etc.

These modes of decay which become more diverse as the energy increases are analogous to the wave guide exits attached to the electromagnetic cavity.

An example of a resonance which can be seen in a number of different exit channels is that of the $\Lambda(1520)$ which has $S = -1$, $I = 0$, $J^P = \frac{3}{2}^-$. In Fig. 47.1 the cross-sections are shown for the reactions

$$\begin{aligned} K^- + p &\to K^- + p \\ &\to \Sigma^+ + \pi^- \\ &\to \Sigma^- + \pi^+ \\ &\to \overline{K^0} + n \\ &\to \Sigma^0 + \pi^0 \\ &\to \Lambda + \pi^+ + \pi^- \end{aligned}$$

as well as the total cross-section for $K^-$ incident on protons. It is seen that the $\Lambda(1520)$ resonance decays into each of the six channels.

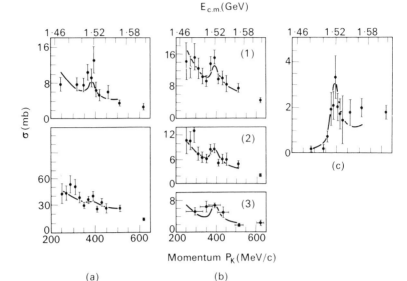

FIGURE 47.1    Evidence of the $\Lambda(1520, J^P = \frac{3}{2}^-)$ resonance in several channels. Cross-sections as a function of momentum for reactions occurring when $K^-$ mesons are incident on protons; (a) $K^-p$ charge exchange and elastic scattering: (1) $\overline{K}^0 n$, (2) $K^- p$; (b) $\Sigma^+ \pi^-$, $\Sigma^- \pi^+$ and $\Sigma^0 \pi^0$ production: (1) $\Sigma^+ \pi^-$, (2) $\Sigma^- \pi^+$, (3) $\Sigma^0 \pi^0$; (c) $\Lambda \pi^+ \pi^-$ production. The solid line corresponds to the best fit of all cross sections, angular distributions, and polarizations (after Tripp, 1966.)

## 48 Nomenclature

The labelling of resonances in particle physics has been rather chaotic, and it has seemed at times that the rate at which resonances are discovered would cause physicists to run out of alphabets.

However, there is a convenient nomenclature for all firmly established strongly-interacting particles, or hadrons, which involves 10 basic symbols. The hadrons can be divided naturally into the mesons and baryons, with baryon numbers 0 and 1 respectively. To characterize completely a particular charge multiplet one needs only the quantum numbers of hypercharge $Y$ and isospin $I$, together with the spin and parity, $J^P$, and the mass. (Strangeness $S$ can be used instead of hypercharge, as $S$ and $Y$ are related by $Y = B + S$ where $B$ is the baryon number.) The symbols assigned to each of the observed combinations of $Y$ and $I$ are listed in Table 48.1.

A particular particle is specified by giving the mass and the spin and parity, $J^P$, in brackets after the basic symbol. For example the

$\rho$-meson and the $\omega$-meson discussed in section 45 are denoted as follows

$$\rho = \pi(765, 1^-)$$
$$\omega = \eta(784, 1^-)$$

The $\rho$ can be regarded as an excited state of the pion, and the $\omega$ can be regarded as an excited state of the $\eta(549, 0^-)$. However in a wider sense, all the mesons can be regarded as the various states of some meson, and all the baryons as the various states of some baryon.

TABLE 48.1    Nomenclature for hadrons

| Mesons | $Y$ | $I$ | Baryons | $Y$ | $I$ |
|--------|-----|-----|---------|-----|-----|
| $\eta$ | 0 | 0 | $N$ | $+1$ | $\frac{1}{2}$ |
| $\pi$ | 0 | 1 | $\Delta$ | $+1$ | $\frac{3}{2}$ |
| $K$ | $+1$ | $\frac{1}{2}$ | $\Lambda$ | 0 | 0 |
| $\overline{K}$ | $-1$ | $\frac{1}{2}$ | $\Sigma$ | 0 | 1 |
| | | | $\Xi$ | $-1$ | $\frac{1}{2}$ |
| | | | $\Omega$ | $-2$ | 0 |

All $\pi$- and $\eta$-mesons are their own antiparticles, so that no symbol is needed for antiparticles of $\pi$- and $\eta$-mesons. For antibaryons, the symbol for the appropriate baryon is used with a bar inserted over the symbol. Some care is needed with this scheme in order to specify a particular charge state unambiguously; we use the convention

$$\overline{\Xi^+} = \overline{\Xi}^-$$
$$\overline{\Sigma^-} = \overline{\Sigma}^+$$

etc.

Note that an antiparticle has the same isospin quantum number $I$ but a hypercharge $Y$ of opposite sign to that of the corresponding particle.

This scheme of nomenclature cannot be used for a particular resonance until the quantum numbers of the resonance have been determined. There is also the disadvantage that the quoted value of the mass of a particle may change as further experimental results become available. Thus the particle specified above as $\pi(765, 1^-)$ is the same particle specified by Chew *et al.* (1964) as $\pi(750, 1^-)$. For these reasons, other symbols are also used for some particles, such as the *R*, *S*, *T*, *U* of Fig. 45.5.

A modification of this scheme of nomenclature distinguishes between mesons of different $G$ parity. The operator $G$ is defined as

$$G = CR_2$$

where $R_2$ denotes a rotation of $\pi$ around the 2-axis in isospin space, and $C$ is the charge conjugation operator. For a particle to be an eigenstate of $G$, it must have $Y=0$ and $B=0$, since $C$ acting on a state of hypercharge $Y$ and baryon number $B$ yields a state with hypercharge $-Y$ and baryon number $-B$, and $Y$ and $B$ are unaltered by rotations in isospin space. Note that the action of $R_2$ is to change $I_3$ to $-I_3$, and the action of $C$ changes $-I_3$ back to $I_3$. Thus the operator $G$ leaves $I_3$ unaltered, and so $G$ and the operator for the third component of isospin commute.

$$[G, I_3] = 0$$

It can be shown (Lee and Yang, 1956) that

$$[G, I_2] = [G, I_1] = 0$$

Thus a state of definite isospin with $Y=0$, $B=0$ can also be an eigenstate of $G$.

A neutral particle with $Y=0$, $B=0$, such as the $\pi^0$, is its own antiparticle, so that if $\psi$ is the state function for such a particle

$$|C\psi|^2 = |\psi|^2$$
$$C\psi = e^{i\delta}\psi$$

Since $C^2 = 1$, then following a similar argument to the treatment of the parity operator in Section 13

$$C\psi = \pm\psi$$

The eigenvalues of $C$ are $\pm 1$. A neutral hadron with $Y=0$, $B=0$ has $I_3 = 0$. We have (Feynman, 1965)

$$R_2\psi_{I, I_3 = 0} = (-1)^I\psi_{I, I_3 = 0}$$

so that for a neutral meson with $Y=0$, $B=0$

$$G = C(-1)^I$$

Since $G$ commutes with the isospin operators, the other mesons in the same isospin multiplet will have the same eigenvalue of $G$. The eigenvalues of $G$ are $\pm 1$. The eigenvalue of $G$ is called the $G$ parity of the state.

The pion has odd $G$ parity

$$G_\pi = -1$$

Using $G$ parity, a modification of the nomenclature in Table 48.1 is used by the Particle Data Group in *Review of Particle Properties* (see end of section 41) for $Y=0$ mesons, as follows

$I = 0$;   $\eta$ if $G$ is even, $\phi$ if $G$ is odd;

$I = 1$;   $\rho$ if $G$ is even, $\pi$ if $G$ is odd

### References

ALFF, C., D. BERLEY, D. COLLEY, N. GELFAND, U. NAUERNBERG, D. MILLER, J. SCHULTZ, J. STEINBERGER, T. H. TAN, H. BRUGGER, P. KRAMER and R. PLANO, *Phys. Rev. Letters* **9** (1952) 322; *Phys. Rev.* **145** (1966) 1072.

ALSTON, M., L. ALVAREZ, P. EBERHARD, M. GOOD, W. GRAZIANO, H. TICHO and S. WOJCICKI, *Phys. Rev. Letters* **5** (1960) 520.

BAREYRE, P., C. BRICMAN and G. VILLET, *Phys. Rev.* **165** (1968) 1730.

BARASHENKOV, V. S., *Interaction Cross Sections of Elementary Particles,* 1968. Israel Program for Scientific Translation, Jerusalem 1968.

BARNES, V. E., P. L. CONNOLLY, D. J. CRENELL, B. B. CULWICK, W. C. DELANEY, W. B. FOWLER, P. E. HAGERTY, E. L. HART, N. HORWITZ, P. V. C. HOUGH, J. E. JENSEN, J. K. KOPP, K. W. LAI, J. LEITNER, J. L. LLOYD, G. W. LONDON, T. W. MORRIS, Y. OREN, R. B. PALMER, A. G. PRODELL, D. RADOJIČIĆ, D. C. RAHM, C. R. RICHARDSON, N. P. SAMIOS, J. R. SANFORD, R. P. SHUTT, J. R. SMITH, D. L. STONEHILL, R. C. STRAND, A. M. THORNDIKE, M. S. WEBSTER, W. J. WILLIS and S. S. YAMAMOTO, *Phys. Rev. Letters* **12** (1964) 204.

BERTANZA, L., V. BRISSON, P. CONNOLLY, E. HART, I. MITTRA, G. MONETI, R. RAU, N. SAMIOS, S. LICHTMAN, I. SKILLICORN, S. YAMAMOTO, L. GRAY, M. GOLDBERG, J. LEITNER and J. WESTGARD, *Phys. Rev. Letters* **9** (1962) 180.

BLIEDEN, H. R., D. FREYTAG, J. GEIBEL, A. R. F. HASSAN, W. KIENZLE, F. LEFÈBRES, B. LEVRAT, B. C. MAGLIĆ, J. SEIGUINOT and A. J. SMITH, *Phys. Letters* **19** (1965) 444; *Nuovo Cimento* **43**A (1966) 71.

CARROLL, L., *Through the Looking-glass,* 1872. Chapter 6.

CENCE, R. J., *Pion–Nucleon Scattering,* 1969. Princeton University Press.

CHEW, G. F., M. GELL-MANN and A. H. ROSENFELD, 'Strongly interacting particles', *Sci. Amer.,* February 1964. (Also available as reprint 296, Freeman, San Francisco.)

CITRON, A., W. GALBRAITH, T. F. KYCIA, B. A. LEONTIC, R. H. PHILLIPS, A. ROUSSET and P. H. SHARP, *Phys. Rev.* **144** (1966) 1101.

DIDDENS, A. N., E. W. JENKINS, T. F. KYCIA and K. F. RILEY, *Phys. Rev. Letters* **10** (1963) 262.

EISBERG, R. M., *Fundamentals of Modern Physics,* 1961. Wiley, New York.

ERWIN, A. R., R. MARCH, W. D. WALKER and E. WEST, *Phys. Rev. Letters* **6** (1961) 628.

FERRO-LUZZI, M., R. D. TRIPP and M. B. WATSON, *Phys. Rev. Letters* **8** (1962) 28.

FEYNMAN, R. P., R. B. LEIGHTON and M. SANDS, *Quantum Mechanics,* Vol. III of *The Feynman Lectures on Physics,* 1965. Addison-Wesley, Reading, Mass. Chapter 18.

FOCACCI, M. N., W. KIENZLE, B. LEVRAT, B. C. MAGLIĆ and M. MARTIN, *Phys. Letters* **17** (1966) 890.

FOWLER, W. B. and SAMIOS, N. P., 'The omega-minus experiment', *Sci. Amer.* October 1964.

HALLIDAY, D., *Introductory Nuclear Physics,* 1950. Wiley, New York.

LEE, T. D. and C. N. YANG, *Nuovo Cimento* **3** (1956) 749.

MAGLIĆ, B. C., L. W. ALVAREZ, A. H. ROSENFELD and M. L. STEVENSON, *Phys. Rev. Letters* **7** (1961) 178.

MAGLIĆ, B. and G. GOSTA, *Phys. Letters* **18** (1965) 185.

PJERROU, G., D. PROWSE, P. SCHLEIN, W. SLATER, D. STORK and H. TICHO, *Phys. Rev. Letters* **9** (1962) 114.

RITTENBERG, A., A. BARBARO-GALTIERI, T. LASINSKI, A. H. ROSENFELD, T. G. TRIPPE, M. ROOS, C. BRICMAN, P. SÖDING, N. BARASH-SCHMIDT, and C. G. WOHL. *Rev. Mod. Phys.* **43** (1971) No. 2, Pt. II.

SCHLEIN, P. E., D. D. CARMONY, G. M. PJERROU, W. E. SLATER, D. H. STORK and H. K. TICHO, *Phys. Rev. Letters* **11** (1963) 167.

SCHÜBELIN, P., *Physics Today* **23** (1970). No. 11, November, 32.

SHAFER, J. B., J. J. MURRAY and D. O. HUWE, *Phys. Rev. Letters* **10** (1963) 179.

SÖDING, P., J. BARTELS, A. BARBARO-GALTIERI, J. E. ENSTROM, T. A. LASINSKI, A. RITTENBERG, A. H. ROSENFELD, T. G. TRIPPE, N. BARASH-SCHMIDT, C. BRICMAN, V. CHALOUPKA, and M. ROOS. *Phys. Letters* **39B** (1972 No. 1.).

TRIPP, R. D. *Baryon Resonances* in Proceedings of the International School of Physics 'Enrico Fermi' Course XXXIII, Strong Interactions, 1966. Academic Press, New York.

WOJCICKI, S. G., *Phys. Rev.* **135B** (1964) 484.

### Exercises

1   The following reactions can contribute to the total cross-section for $\pi^- p$. Calculate the kinetic energy of the pion in the laboratory frame, at threshold for each reaction

$$\pi^- + p \rightarrow \pi^- + \pi^0 + p$$
$$\pi^- + p \rightarrow K^0 + \Lambda^0$$
$$\pi^- + p \rightarrow \Sigma^- + K^+$$
$$\pi^- + p \rightarrow \pi^- + \pi^- + \pi^+ + p$$

2   In the case of the scattering of pions by protons, calculate and graph the momentum $p_\pi$ and the kinetic energy $T_\pi$ of the pion in the laboratory frame as a function of $E$, the total energy of the pion–proton system in the centre-of-mass frame. Express $T_\pi$ and $E$ in GeV, and $p_\pi$ in GeV/$c$. Using the graph, find $p_\pi$ and $T_\pi$ for the following resonances,

$$N(1520), \quad N(2650), \quad \Delta(1236), \quad \Delta(2850)$$

3   What isospin states are involved in the following reactions?

$$\pi^+ + p \rightarrow \pi^+ + p$$
$$\pi^+ + p \rightarrow \pi^+ + \pi^+ + n$$
$$\pi^- + p \rightarrow \pi^- + p$$
$$\pi^- + p \rightarrow K^0 + \Lambda^0$$

4   Show that

$$\sigma_{\frac{1}{2}} = \tfrac{3}{2}\sigma_{\text{tot}}^- - \tfrac{1}{2}\sigma_{\text{tot}}^+$$

where $\sigma_{tot}^-$, $\sigma_{tot}^+$ are the total cross-sections for $\pi^-p$ and $\pi^+p$ respectively, and $\sigma_{\frac{1}{2}}$ is the $I=\frac{1}{2}$ pion–nucleon cross-section.

5    Figure 45.6 (from Focacci *et al.*, 1966) shows that the plot of $M_x^2$ against peak number is a straight line. Make a similar plot, including the additional peaks from Fig. 45.5. Can you draw any conclusions?

# *SU* (3) multiplets of hadrons

# 11

## 49 Introduction

We have seen that the strongly interacting particles, or hadrons, characterized by their masses, $J^P, B, I, I_3$ and $Y$ values, can be grouped naturally into charge multiplets, within which the $I_3$ values vary from $-I$ to $+I$ but all the other characteristics are the same, except for small mass differences.

For instance the nucleon is an isospin doublet with $J^P = \frac{1}{2}^+$, $B=1$, $I = \frac{1}{2}$, $Y=1$, with $I_3 = +\frac{1}{2}$ for the proton and $I_3 = -\frac{1}{2}$ for the neutron, and the masses are

$$M_p = 938 \cdot 2 \text{ MeV}$$
$$M_n = 939 \cdot 6 \text{ MeV}$$

There is a mass difference of

$$M_n - M_p = 1 \cdot 4 \text{ MeV}$$

which is small compared to the mean mass of the nucleon

$$M_N = \tfrac{1}{2}(M_p + M_n) = 939 \text{ MeV}$$

It is possible to arrange the hadrons into even larger multiplets in which $Y$, $I$ and $I_3$ vary, but where $J^P$ and $B$ are the same for all members of each multiplet. For these larger multiplets, there are larger differences in mass between members of a multiplet.

For instance, let us consider the baryons with $J^P = \frac{1}{2}^+$ and with low masses. We have already mentioned the nucleon. The next heavier $\frac{1}{2}^+$ baryon is the $\Lambda^0$, with $Y=0$, $I=0$, $I_3=0$ and

$$M_{\Lambda^0} = 1115 \cdot 6 \text{ MeV}$$

so that

$$\begin{aligned} M_\Lambda - M_N &= 1116 - 939 \\ &= 177 \text{ MeV} \end{aligned}$$

Next we come to $\Sigma^-$, $\Sigma^0$, $\Sigma^+$ with $I_3 = -1, 0, +1$ respectively and $Y = 0$, $I = 1$ and masses:

$$\left.\begin{array}{l} M_{\Sigma^-} = 1197\cdot3 \text{ MeV} \\ M_{\Sigma^0} = 1192\cdot5 \text{ MeV} \\ M_{\Sigma^+} = 1189\cdot5 \text{ MeV} \end{array}\right\} M_\Sigma = 1193 \text{ MeV}$$

Compare

$$M_{\Sigma^-} - M_{\Sigma^0} = 4\cdot8 \text{ MeV}$$
$$M_{\Sigma^0} - M_{\Sigma^+} = 3\cdot0 \text{ MeV}$$

with

$$M_\Sigma - M_\Lambda = 77 \text{ MeV}$$

The next $J^P = \frac{1}{2}^+$ baryons are $\Xi^-$ and $\Xi^0$ with $I_3 = -\frac{1}{2}, +\frac{1}{2}$ respectively and $Y = -1$, $I = \frac{1}{2}$, and masses:

$$\left.\begin{array}{l} M_{\Xi^-} = 1321 \text{ MeV} \\ M_{\Xi^0} = 1315 \text{ MeV} \end{array}\right\} M_\Xi = 1318 \text{ MeV}$$

Compare

$$M_{\Xi^-} - M_{\Xi^0} = 6 \text{ MeV}$$

with

$$M_\Xi - M_\Sigma = 125 \text{ MeV}$$

The arrangement of $N$, $\Lambda$, $\Sigma$ and $\Xi$ into one multiplet, an octet since it contains eight particles, is illustrated in Fig. 49.1. The mass splittings between the different isospin multiplets in the octet are about 20 times the mass splitting within isospin multiplets.

The existence of this octet, and of other multiplets of hadrons, each containing several isospin multiplets, leads us to imagine that there is some 'superstrong' interaction, which by itself would yield the same mass for all particles in the large multiplet. The mass splitting is attributed to some symmetry-breaking perturbation which depends on $Y$ and $I$. The symmetry breaking perturbation is however part of the strong interaction and so causes large mass splittings.

The $\Omega^-$ and baryon resonances with $J^P = \frac{3}{2}^+$ can be arranged into a multiplet of 10 particles, a decuplet, as discussed in Section 44, and shown again in Fig. 49.2.

Note the threefold symmetry of the diagrams for both $J^P = \frac{1}{2}^+$ baryons, Fig. 49.1, and $J^P = \frac{3}{2}^+$ baryons, Fig. 49.2.

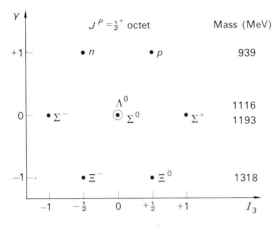

FIGURE 49.1   The octet of $J^P = \frac{1}{2}^+$ baryons. In these diagrams, an additional circle is used wherever there is another particle with the same $Y$ and $I_3$.

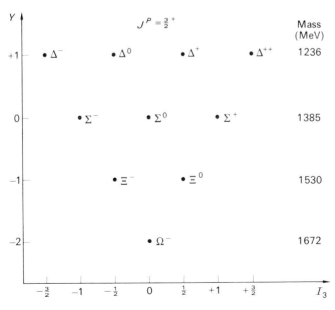

FIGURE 49.2   The decuplet of $J^P = \frac{3}{2}^+$ baryons.

In 1961, M. Gell-Mann and Y. Ne'eman, working independently, related the symmetries of hadrons to the properties of the group $SU(3)$. The group $SU(3)$ is the group of all unimodular unitary $3 \times 3$ matrices, i.e. the group of all $3 \times 3$ matrices $U$ such that $U$ is unitary

$$U^\dagger = U^{-1}$$

and unimodular,

$$\text{determinant } U = 1$$

$U^\dagger$, the hermitian adjoint of $U$, is obtained by interchanging rows and columns and taking the complex conjugate of each element, i.e.

$$(U^\dagger)_{ij} = U^*_{ji}$$

$SU$ stands for 'special unitary', the term 'special' referring to the unimodularity condition. We shall not need the above definition of $SU(3)$ nor any knowledge of the mathematics of group theory for the applications to particle physics discussed in this book. However, some background of group theory is given in the next section.

Gell-Mann called the analysis of the hadrons by $SU(3)$ the 'eightfold way' because there are eight quantities of interest in the algebra of this group and also because it recalls a saying attributed to Buddha:

> 'Now this, O monks, is noble truth that leads to the cessation of pain: this is the noble *Eightfold Way*: namely, right views, right intention, right speech, right action, right living, right effort, right mindfulness, right concentration.'

(Quoted in Chew *et al*, 1964.)

The three components of isospin together with the hypercharge constitute four of the eight quantities of the eightfold way of hadron physics. The other four cannot be specified so easily.

The application of $SU(3)$ to particle physics is now usually referred to as unitary symmetry.

The prediction of the $\Omega^-$, completing the decuplet of Fig. 49.2, was one of the early successes of the eightfold way.

## 50 Group theory in physics

Group theory is important in physics since group theory is the mathematics of symmetry. In this respect, it is interesting to read Weyl (1952).

*Definition of a group*

A group is a set of elements with some form of product which associates with any two elements of the group a third element also belonging to the group. The product is associative. The group contains an identity and each element of the group possesses an inverse also belonging to the group. i.e.

(a) Given elements $a$ and $b$ in a group $G$, $ab$ is also in $G$
(b) The associative law holds, i.e. $a(bc)=(ab)c$ for all $a,b,c$ in $G$
(c) There is an identity (or unit element) $e$ such that $ea=ae=a$ for every $a$ in $G$
(d) For every $a$ in $G$ there is an element $a^{-1}$ in $G$ such that $a^{-1}a = aa^{-1} = e$.

For example, the rotations about an axis, say the $z$ axis, form a group. The product of two rotations is one rotation followed by the other. The identity is to do nothing. The inverse of a rotation through $\theta$ is a rotation through $-\theta$. The set of all rotations in three-dimensional space forms a group, the three-dimensional rotation group $O(3)$. Similarly translations in three-dimensional space form a group. The product of two translations, one translation followed by the other, is also a translation.

The above examples are continuous groups. A continuous group is a group whose elements can be characterized by parameters varying continuously in a certain region. For instance, rotations about an axis are characterized by the angle $\theta$ varying from 0 to $2\pi$. Translations in three-dimensional space can be characterized by the three components of the displacement $x,y,z$ of the origin. $x,y,z$ each vary from $-\infty$ to $+\infty$.

There are also groups containing a finite number of elements. For instance, $e$, the identity and $P$, inversion through the origin form a group of two elements

$$eP = P = Pe$$
$$PP = e$$

The set of transformations which leave the Schrödinger equation of a system invariant form a group, called the symmetry group of that system. For if the transformations $A$ and $B$ separately leave the Schrödinger equation invariant, the product of the two transformations also leaves the Schrödinger equation invariant.

The symmetry group of a quantum-mechanical system determines the multiplicities that can occur for the levels of that system. For instance, for the three-dimensional rotation group $O(3)$, the multiplicities

$$1, 3, 5, 7, \text{etc.}$$

occur, corresponding to the spins

$$0, 1, 2, 3, \text{etc.}$$

respectively. For the group $SU(2)$, the group of $2 \times 2$ unitary matrices with unit determinant, all multiplicities can occur. The group $O(3)$ is related to the group $SU(2)$, and it is as a consequence of this relation that half-integer spins occur in quantum mechanics, and it is possible to build up any spin by adding together enough terms, each of spin $\frac{1}{2}$. For example, both spin 0 and 1 can be obtained by adding together two terms, each of spin $\frac{1}{2}$

$$\tfrac{1}{2} + \tfrac{1}{2} = \mathbf{0} \text{ or } \mathbf{1}$$

Spin $\frac{1}{2}$ corresponds to a doublet, there being two spin states. $2 \times 2$ matrices can be defined for transformations of these two states, giving the group $SU(2)$. The various spin multiplets, singlets, doublets, triplets, etc., can be built up by compounding a sufficient number of doublets.

The elements of the group $SU(3)$ are $3 \times 3$ matrices and can be defined using three states, a triplet. The various $SU(3)$ multiplets can be built up by compounding triplets, as we shall see in detail later for certain multiplets. The $SU(3)$ multiplets of concern for hadrons are singlets, octets and decuplets.

### 51  *SU*(3) classification of baryons and mesons

In Section 50, we have already encountered an octet, that of $J^P = \frac{1}{2}^+$ baryons, and a decuplet, that of $J^P = \frac{3}{2}^+$ baryons. An example of an $SU(3)$ singlet is the $\Lambda(1405, \frac{1}{2}^-)$ which has $Y=0$ and $I=0$.

The masses of different members of a $SU(3)$ multiplet are not the same, see Figs. 49.1 and 49.2, and so the $SU(3)$ symmetry is not exact but must be broken. Assuming that $SU(3)$ symmetry is broken in a particularly simple way, by using group theory, Okubo derived the following formula for the masses within an $SU(3)$ multiplet

$$M = M_0 + aY + b[I(I+1) - \tfrac{1}{4}Y^2] \qquad (51.1)$$

$M_0$, $a$ and $b$ should be constant within a given multiplet. Equation (51.1) is called the Gell-Mann–Okubo mass formula. In this form, equation (51.1), the formula applies to baryons. For the mesons, better agreement with experiment is obtained if $M$ is replaced by $M^2$ in equation (51.1).

For an octet of baryons, equation (51.1) yields

$$\tfrac{1}{2}(M_N + M_\Xi) = \tfrac{1}{4}(3M_\Lambda + M_\Sigma) \tag{51.2}$$

For the octet of $J^P = \tfrac{1}{2}^+$ baryons, the experimental values of the masses are in good agreement with equation (51.2). (See Exercise 1.)

For the decuplet of $J^P = \tfrac{3}{2}^+$ baryons the agreement with the Gell-Mann–Okubo mass formula is again excellent. For the decuplet (see Fig. 49.2),

$$I = 1 + \tfrac{1}{2}Y \tag{51.3}$$

and equation (51.1) becomes

$$M = (M_0 + 2b) + (a + \tfrac{3}{2}b)Y \tag{51.4}$$

So that the Gell-Mann–Okubo mass formula predicts equal spacing of the masses of the decuplet,

$$M_\Sigma - M_\Lambda = M_\Xi - M_\Sigma = M_\Omega - M_\Xi \tag{51.5}$$

in agreement with the observed masses. In fact the mass of the $\Omega^-$ was predicted quite accurately by the Gell-Mann–Okubo mass formula before this particle was discovered.

Other baryon resonances can also be arranged in $SU(3)$ multiplets. For instance, the *Review of Particle Properties* (Söding *et al.*, 1972) lists the following $J^P = \tfrac{3}{2}^-$ baryons

$$N(1520, \tfrac{3}{2}^-)$$
$$\Lambda(1518, \tfrac{3}{2}^-) \text{ and } \Lambda(1690, \tfrac{3}{2}^-)$$
$$\Sigma(1670, \tfrac{3}{2}^-)$$

Taking the $\Lambda(1518, \tfrac{3}{2}^-)$ as a member of the octet, the Gell-Mann–Okubo mass formula, equation (51.4), predicts $\Xi(1582, \tfrac{3}{2}^-)$, but no resonance has been observed that can be identified with this prediction. On the other hand, taking the $\Lambda(1690, \tfrac{3}{2}^-)$ as a member of the octet, equation (51.4) predicts $\Xi(1840, \tfrac{3}{2}^-)$. The *Review of Particle Properties* (Söding *et al.*, 1972) lists a $\Xi$ resonance with mass of 1795 to 1870 MeV and undetermined spin and parity which can be identified with the predicted $\Xi(1840, \tfrac{3}{2}^-)$. The $SU(3)$ scheme thus predicts the spin and parity of this resonance. The conjectured $J^P = \tfrac{3}{2}^-$ baryon octet is shown in Fig. 51.1. The $\Lambda(1518, \tfrac{3}{2}^-)$ is then identified as an $SU(3)$ singlet.

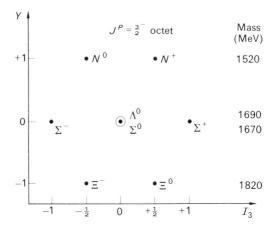

FIGURE 51.1  $J^P = \frac{3}{2}^-$ baryons. For $\Xi$, $J^P$ is not yet established and the mass is uncertain.

Table 51.1 shows the identification of some of the observed baryon resonances as members of $SU(3)$ multiplets. The $SU(3)$ scheme makes many predictions about spins and parities, and also predicts the masses of resonances to complete multiplets by using the Gell-Mann–Okubo mass formula.

TABLE 51.1   Possible $SU(3)$ multiplets of baryons. The mass of states for which $J^P$ is not established are placed in brackets

### Singlets

| $J^P$ | $\Lambda$ |
|---|---|
| $\frac{1}{2}^-$ | 1405 |
| $\frac{3}{2}^-$ | 1518 |
| $\frac{7}{2}^-$ | 2100 |

### Octets

| $J^P$ | $N$ | $\Lambda$ | $\Sigma$ | $\Xi$ |
|---|---|---|---|---|
| $\frac{1}{2}^+$ | 939 | 1116 | 1193 | 1318 |
| $\frac{1}{2}^+$ | 1470 | (1750) | (1620) | — |
| $\frac{3}{2}^-$ | 1520 | 1690 | 1670 | (1820) |
| $\frac{1}{2}^-$ | 1535 | 1670 | 1750 | — |
| $\frac{5}{2}^-$ | 1670 | 1830 | 1765 | (1930) |
| $\frac{5}{2}^+$ | 1688 | 1815 | 1905 | (2030) |

### Decuplets

| $J^P$ | $\Delta$ | $\Sigma$ | $\Xi$ | $\Omega$ |
|---|---|---|---|---|
| $\frac{3}{2}^+$ | 1236 | 1385 | 1530 | 1672 |
| $\frac{7}{2}^+$ | 1950 | 2030 | — | — |

We expect all members of an $SU(3)$ multiplet to have the same quantum numbers other than $Y$, $I$, $I_3$ which are $SU(3)$ quantum numbers. For instance the baryon number $B$ is the same throughout a multiplet. Thus, to every baryon multiplet with $B=1$, there will be a corresponding multiplet of antibaryons with $B=-1$. For baryons, particles and antiparticles are in separate $SU(3)$ multiplets. The mesons have $B=0$, and so a meson multiplet can contain both particles and antiparticles. For instance, the $\pi^-$ is the antiparticle of the $\pi^+$, and both are in the $I=1$ isospin triplet. So we can expect $K^0$, $K^+$, $K^-$ and $\overline{K^0}$ within the same $SU(3)$ multiplets.

The pseudoscalar mesons listed in Table 51.2 can be arranged as an $SU(3)$ octet and an $SU(3)$ singlet as shown in Fig. 51.2. For meson octets, masses for $Y=+1$ and $Y=-1$ are the same, since these are antiparticles of each other.

TABLE 51.2    $J^P = 0^-$-mesons

|  | Mass (MeV) | $Y$ | $I$ |
|---|---|---|---|
| $\pi^+, \pi^-$ <br> $\pi^0$ | 139·6 <br> 135·0 | 0 | 1 |
| $K^+$ <br> $K^-$ | 493·8 | $+1$ <br> $-1$ | $\frac{1}{2}$ <br> $\frac{1}{2}$ |
| $K^0$ <br> $\overline{K^0}$ | 497·8 | $+1$ <br> $-1$ | $\frac{1}{2}$ <br> $\frac{1}{2}$ |
| $\eta$ | 549 | 0 | 0 |
| $\eta'$ | 958 | 0 | 0 |

Applying the Gell-Mann–Okubo mass formula, equation (51.1), to the squares of meson masses, we obtain

$$M_K^2 = (3M_\eta^2 + M_\pi^2)/4 \tag{51.6}$$

which is in good agreement with the experimental results. The same formula linear in the masses does not agree with the observed masses.

The vector mesons, $J^P = 1^-$,

$$\pi(765, 1^-) = \rho$$
$$K(892, 1^-) = K^*$$
$$\eta(784, 1^-) = \omega$$
$$\eta(1019, 1^-) = \phi$$

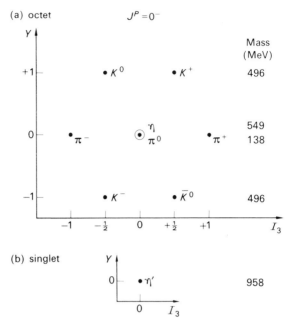

FIGURE 51.2   $SU(3)$ octet and singlet of pseudoscalar mesons $J^P = 0^-$.

can also be arranged as an octet and singlet as shown in Fig. 51.3. However, the Gell-Mann–Okubo mass formula does not seem to hold for the $1^-$ mesons, regardless of whether the masses or the squares of the masses are used. The Gell-Mann–Okubo mass formula for the squares of meson masses can be written

$$M_\eta^2 = (4M_K^2 - M_\pi^2)/3$$

Using the observed masses of the $K^*$ and the $\rho$ yields

$$M_\eta^2 = 0.87 \ (\text{GeV})^2$$

which falls between the experimental values of

$$M_\eta^2(784) = 0.61 \ (\text{GeV})^2$$
$$M_\eta^2(1019) = 1.04 \ (\text{GeV})^2$$

The $\eta$ of the $SU(3)$ octet can be regarded as a linear superposition of the $\eta(784)$ and the $\eta(1019)$. The $1^-$-mesons can better be regarded as a nonet, rather than a separate octet and singlet. The $\eta(784)$ and $\eta(1019)$ can be regarded as two orthogonal superpositions of an octet $\eta$ and a singlet $\eta$. In the following sections, we shall encounter a

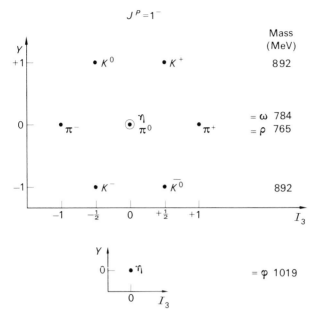

FIGURE 51.3  *SU*(3) octet and singlet of vector mesons, $J^P = 1^-$.

TABLE 51.3  Classification of mesons into nonets. (Symbols in table entries show the colloquial names of the mesons) (Masses in MeV)

| $J^P$ | $I = 1$ <br> $\pi$ | $I = \frac{1}{2}$ <br> $K$ | $I = 0$ <br> $\eta$ |
|---|---|---|---|
| $0^-$ | $\pi$, 138 | $K$, 496 | $\eta$, 549 <br> $\eta'$, 958 |
| $1^-$ | $\rho$, 765 | 892 | $\omega$, 784 <br> $\phi$, 1019 |
| $0^+$ | $\delta$, 966 | — | 700 <br> $S^*$, 1060 |
| $1^+$ | $A1$, 1070 | 1240 | $D$, 1288 <br> — |
| $2^+$ | $A2$, 1300 | 1420 | $f$, 1260 <br> $f'$, 1514 |
| $1^+$ | $B$, 1235 | — | — <br> — |

model of mesons which leads us to expect mesons to occur in nonets.

Table 51.3 shows the classification of mesons into nonets.

See Gell-Mann and Ne'eman (1964) for an anthology of papers on $SU(3)$, and Lipkin (1966) for an account of the theory of $SU(3)$ without the impediments of group theory.

## 52 The quark model

We have seen that the hadrons can be classified according to a three-fold symmetry described by the group $SU(3)$. The mass splitting within a multiplet can to some extent be accounted for by assuming a certain kind of symmetry breaking perturbation.

The hadrons can be arranged in multiplets, the multiplicities of which can be derived from group theory. However, not all the multiplicities predicted by $SU(3)$ symmetry occur among the hadrons. This is perhaps not too surprising if we consider a similar case in the theory of ordinary spin, in which the allowed multiplicities are 1, 2, 3, 4, 5, etc., corresponding to spins $0, \frac{1}{2}, 1, \frac{3}{2}, 2$, etc. But, for any given system, either only integer or only half-odd-integer spins can occur, so that the multiplicities for any particular system are either all even or all odd. There are also other systems in physics which have some symmetry group, and not all the multiplicities that are allowed by group theory occur for the system. It is not unreasonable that there are additional conditions on multiplets for elementary particles which restrict the multiplets to the ones which are found in nature.

In particular, the $SU(3)$ triplets do not seem to occur for hadrons. The triplet is the fundamental multiplet for the group $SU(3)$. All other $SU(3)$ multiplets can be formed by combining triplets. Gell-Mann (1964) and, independently, Zweig (1965) have postulated the existence of an $SU(3)$ triplet of hypothetical particles, from which all the hadrons can be constructed. These hypothetical particles were called 'aces' by Zweig and 'quarks' by Gell-Mann. 'Quark' is a word used by James Joyce (1939) in *Finnegan's Wake*.

Let us designate the three quarks by $a,b,c$. The quarks each have spin $\frac{1}{2}$, and the other quantum numbers are as shown in Table 52.1. $a$ and $b$ form an isospin doublet and $c$ is an isospin singlet. The triplet of quarks is depicted in Fig. 52.1a, and this triplet is denoted by 3. The antiparticles of the quarks form another triplet depicted in Fig. 52.1b and denoted by $\bar{3}$. The diagram for the $\bar{3}$ is obtained from the diagram for the 3 by reflection in the origin.

Baryons are composed of three quarks and so have half-odd-integer spin. All members of an $SU(3)$ multiplet have the same baryon

TABLE 52.1 Quantum numbers of quarks

| Quark | $I_3$ | $I$ | $Y$ |
|-------|-------|-----|-----|
| $a$ | $+\frac{1}{2}$ | $\frac{1}{2}$ | $+\frac{1}{3}$ |
| $b$ | $-\frac{1}{2}$ | | |
| $c$ | $0$ | $0$ | $-\frac{2}{3}$ |

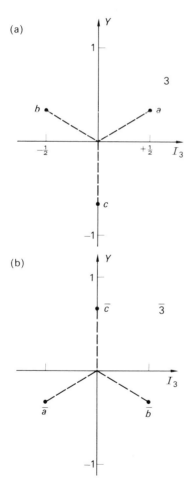

FIGURE 52.1 Diagrams for triplets, 3 of quarks and $\bar{3}$ of antiquarks.

number $B$ and so since baryons have $B=1$, quarks must have $B=\frac{1}{3}$. The antiquarks then have $B=-\frac{1}{3}$. Mesons have $B=0$ and are made up of a quark and an antiquark.

The quark model has been reviewed recently by Morpurgo (1970), and an anthology on the quark model is provided by Kokkedee (1969).

## 53  The quark model of mesons

Since quark and antiquark each have spin $\frac{1}{2}$, a state of a quark and antiquark with no orbital angular momentum, i.e. an $s$ state with $l=0$, will have spin 0 or spin 1. For a state of two particles of the same intrinsic parity with $l=0$, the parity is even. From the Dirac theory of spin-$\frac{1}{2}$ particles, a state of a particle and its antiparticle with $l=0$ has odd parity. This can be regarded as particle and anti-particle having the opposite intrinsic parity. The states of a quark and antiquark with $l=0$ will have spins and parities $J^P=0^-$ and $1^-$, and we have seen that mesons occur with $J^P=0^-$ and $1^-$.

In general, for a state of quark and antiquark with orbital angular momentum quantum number $l$ the parity is $(-1)^{l+1}$. For a state of quark and antiquark with $l=1$ ($p$-state), the parity is even and the possible values of the total angular momentum $J$ are obtained by first compounding the spin of the quark and antiquark to give $S=0$ or $S=1$, and then adding $l=1$ to give

$$J^P = 1^+, 0^+, 1^+, 2^+$$

The nonet of $2^+$-mesons can be described as $p$-states of quark and antiquark. Some mesons have been observed with the other values of $J^P$ corresponding to $p$ states of quark–antiquark.

Diagrams such as those in Figs. 52.1, 51.3, 51.2, 51.1 are called weight diagrams since $I_3$ and $Y$ correspond to quantities called weights in group theory. Obtaining the higher multiplets of $SU(3)$, by compounding numbers of quarks and antiquarks, can be represented by superimposing weight diagrams for 3 and $\bar{3}$. For instance, the combinations of one quark and one antiquark can be determined by placing the origin of the antiquark weight diagram on each quark, in turn, of the weight diagram of the 3. This is shown in Fig. 53.1. Nine states are formed, and these form an $SU(3)$ octet and singlet. Symbolically this is written as

$$3 \otimes \bar{3} = 8 \oplus 1 \tag{53.1}$$

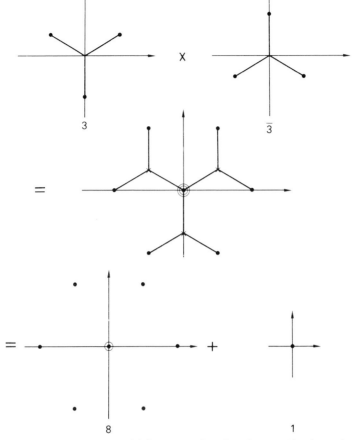

FIGURE 53.1   *SU*(3) multiplets occurring for the combination of quark and antiquark. The × in the second stage indicate the positions of the quarks on which the diagram for $\bar{3}$ is superimposed.

## 54  Properties of quarks

The quarks can be regarded as a convenient mathematical device for finding the consequences of *SU*(3) symmetry. It is not absolutely necessary to regard them as having a physical existence. However, if quarks exist, their properties are very remarkable. Using $Q = I_3 + Y/2$, the electric charges of the quarks are as follows

$$
\begin{array}{llll}
a & +\tfrac{2}{3} & \bar{a} & -\tfrac{2}{3} \\
b & -\tfrac{1}{3} & \bar{b} & +\tfrac{1}{3} \\
c & -\tfrac{1}{3} & \bar{c} & +\tfrac{1}{3}
\end{array}
$$

Conservation of charge and baryon number would ensure that at least one quark would be absolutely stable, $q_1$ the quark with the lowest mass. The other two quarks would decay into the stable quark by weak decays such as

$$q_2 \rightarrow q_1 + \mu + \nu$$
$$q_3 \rightarrow q_1 + \mu + \nu$$

If quarks exist, we might expect to find stable quarks and antiquarks produced in pairs by cosmic rays or by particles from high energy accelerators. Morpurgo (1970) has reviewed the search for quarks. So far, high energy accelerator experiments have given no indication of the existence of quarks. Experiments using the Serpukhov 70 GeV accelerator have established that the total cross-section for the production of quarks with mass $\leqslant 5 \text{GeV}$ is less than $4 \times 10^{-37}$ cm$^2$ for the production of quarks with charge $Q = \frac{2}{3}$ (Antipov *et al.*, 1970a) and less than $3 \times 10^{-39}$ cm$^2$ for the production of quarks with charge $Q = -\frac{1}{3}$ (Antipov *et al.*, 1970b). The negative results of experiments on quark production can be explained by assuming that the quarks have extremely large masses and so require more energy than is available from accelerators. But this assumption of large quark mass brings with it another unfamiliar situation. In order that quark–antiquark pairs will have the correct meson masses, which are comparatively small, the binding energy must be comparable with the sum of the masses of the constituent particles. Usually in atomic and nuclear physics, the mass of a composite bound system is only slightly less than the sum of the masses of the constituents. For instance, the binding energy of a deuteron is very much smaller than the rest mass of the nucleon. Physicists have had no previous experience of systems with binding energies comparable to rest masses, and an extremely relativistic theory is probably necessary to deal adequately with such a situation. An adequate relativistic quantum theory is not yet available.

There are very few cosmic-ray particles with very high energies, and so the production rate of quarks by cosmic rays would be small. Many experiments with cosmic rays have given no indication of quarks. McCusker and Cairns (1969) and Cairns *et al.* (1969) have observed cloud chamber tracks which they interpret as tracks of quarks associated with very energetic cosmic-ray showers. However, this interpretation of the cloud chamber tracks is not generally accepted and has been criticized by several workers (e.g. Adair and Kasha, 1969). This experiment has been repeated with greater sensitivity (Clark et al. 1971) and no evidence for quarks was found.

## 55 Baryons

The baryons can be considered to be bound states of three quarks. Since there are three possible states $a$, $b$, $c$ for each quark, for a combination of three quarks there are $3 \times 3 \times 3 = 27$ states. We wish to know what multiplets are obtained by combining three triplets.

For $SU(2)$, which is the group for spin or isospin, multiplicities can be obtained from the vector model. For instance, combining two systems of multiplicity $(2j_1 + 1)$ and $(2j_2 + 1)$ yields multiplicities

$$\{2(j_1 + j_2) + 1)\}, \{2(j_1 + j_2) - 1)\}, \ldots, \{2|j_1 - j_2| + 1\}$$

i.e. $(2j + 1)$ where $j$ runs from $j_1 + j_2$ to $|j_1 - j_2|$ in steps of 1. For combining the isospins of quarks, the vector model can be used, but the rest of the problem is more difficult. The multiplicities can be

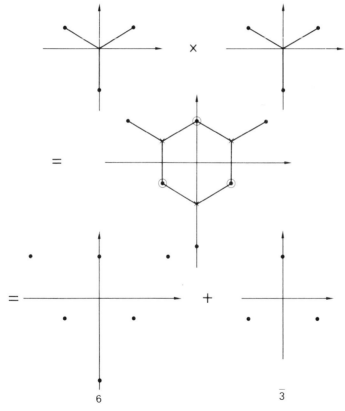

FIGURE 55.1    Illustrating the combination of two quarks

$$3 \otimes 3 = 6 \oplus \bar{3}.$$

determined by group theory which is beyond the scope of this book. (See Lichtenberg, 1970.) For the combination of three triplets, the answer is given symbolically by

$$3 \otimes 3 \otimes 3 = 1 \oplus 8 \oplus 8 \oplus 10 \qquad (55.1)$$

The 27 states of three quarks form a singlet, two octets and a decuplet. These are just the multiplets observed for baryons.

In equation (55.1), the octet occurs twice. This is analogous in the case of spins that in adding three spin-$\frac{1}{2}$ terms,

$$\tfrac{1}{2} + \tfrac{1}{2} + \tfrac{1}{2}$$

spin $\frac{1}{2}$ occurs twice. Writing

$$\mathbf{S}_{12} = \mathbf{s}_1 + \mathbf{s}_2$$
$$\mathbf{S} = \mathbf{S}_{12} + \mathbf{s}_3$$

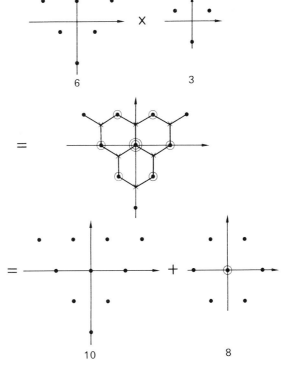

FIGURE 55.2   $6 \otimes 3 = 10 \oplus 8$.

where $s_1, s_2, s_3 = \frac{1}{2}$, then we have

$$S_{12} = 0 \quad \text{so that } S = \frac{1}{2}$$

or

$$S_{12} = 1 \quad \text{so that } S = \frac{1}{2} \text{ or } \frac{3}{2}$$

There are two different ways of having $S = \frac{1}{2}$, namely with $S_{12} = 0$ or $S_{12} = 1$.

The combination of three quarks to form the multiplets of equation (55.1) can be shown pictorially in the same manner used in Fig. 53.1 for combining quark and antiquark. First consider the states of two quarks as shown in Fig. 55.1. This can be written as

$$3 \otimes 3 = 6 \oplus \bar{3} \tag{55.2}$$

This sextet and triplet have non-integral charges and baryon numbers, and so do not occur among the physical hadrons. We now add a third quark to the 6 and $\bar{3}$ in turn. Figure 55.2 shows

$$6 \otimes 3 = 10 \oplus 8 \tag{55.3}$$

In the same manner as in the treatment of mesons, Fig. 53.1,

$$\bar{3} \otimes 3 = 1 \oplus 8 \tag{55.4}$$

Finally combining equations (55.4), (55.3) and (55.2),

$$3 \otimes 3 \otimes 3 = (6 \oplus \bar{3}) \otimes 3$$
$$= 1 \oplus 8 \oplus 8 \oplus 10 \tag{55.5}$$

## 56 Mass splitting in the meson multiplets

If $SU(3)$ symmetry were exact for hadrons, all particles in an $SU(3)$ multiplet would have the same mass. As this is not the case, the $SU(3)$ symmetry cannot be exact.

A simple way to break the $SU(3)$ symmetry is to postulate that the isospin-singlet quark $c$ has a mass $m + \Delta$ different from the mass $m$ of the isospin-doublet quarks $a$ and $b$. To preserve isospin invariance, quarks $a$ and $b$ must have the same mass, apart from electromagnetic corrections.

Let us consider a simple model in which the interaction between quark and antiquark is represented by a very deep square-well potential of depth $V$. We assume that $V$ and the range of the potential are sufficiently large so that the kinetic energy $T$ of the system is independent of small deviations of the mass of one or both of the particles from the value $m$.

The depth and range of the potential is assumed to be independent of the isospin and hypercharge and to depend only on the spin state of the quark–antiquark pair. In general the potential will be different for $S=0$ and $S=1$ where $S$ is the spin quantum number for the sum of the spins of the quark and antiquark,

$$\mathbf{S} = \mathbf{s}_q + \mathbf{s}_{\bar{q}}$$

With the above assumptions, $V$ and $T$ are constant throughout a meson nonet, as is also the binding energy

$$\lambda c^2 = V - T$$

The mass of the system of quark and antiquark, and so the mass of the meson, is

$$M = M_q + M_{\bar{q}} - \lambda \tag{56.1}$$

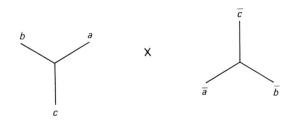

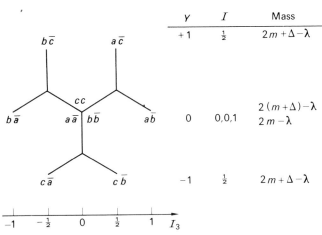

FIGURE 56.1   Nonet of mesons made up of quark and antiquark. $a$ and $b$ have mass $m$. $c$ has mass $m + \Delta$.

The states of a nonet, with their masses, are shown in Fig. 56.1. States $b\bar{c}$, $a\bar{c}$ are obtained by adding isospin $\frac{1}{2}$, $a$ and $b$, to isospin 0, $c$, and so have isospin $I=\frac{1}{2}$, and so have the quantum numbers of kaons. Similarly $c\bar{a}$, $c\bar{b}$ are the isospin doublet $K^-$ and $\bar{K}^0$. $c\bar{c}$ has $I=0$, since $I=0$ for both $c$ and $\bar{c}$. The combination of two isospin doublets $a$, $b$ and $\bar{a}$, $\bar{b}$ yields isospin $I=0$ and $I=1$. $a\bar{b}$ and $b\bar{a}$ are the $I=1$ states with $I_3=+1$ and $I_3=-1$ respectively. There will be two orthogonal linear superpositions of $a\bar{a}$ and $b\bar{b}$, one with $I=0$ and one with $I=1$.

On this simple model, within each meson nonet, we expect the pattern of masses shown in Fig. 56.2a. As shown in Fig. 56.2b, the observed masses of the $1^-$ mesons are in as good agreement with this prediction as can reasonably be expected for such a crude model. As seen from Table 51.3, the observed masses of the $0^-$ mesons

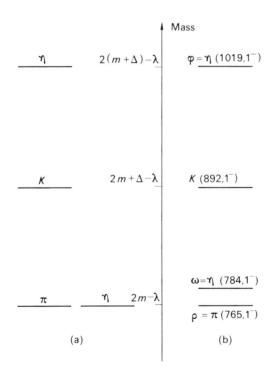

FIGURE 56.2 (a) Pattern of masses expected for a nonet of mesons. (b) Masses of nonet of $1^-$ mesons. The scale is chosen so that $\rho=\pi(765, 1^-)$ and $K(892, 1^-)$ are in the predicted positions.

clearly do not fit this pattern. However, there are two $I=0$, $Y=0$ states in the nonet, and the observed $\eta$ particles can be linear superpositions of these two states; i.e. mixing of the two $I=0$ states can occur. Neither the $I=1$ or the $I=\frac{1}{2}$ states have any states to mix with, and writing the mass as $M^I$ for isospin $I$, the model predicts

$$M_{\frac{1}{2}} - M_1 = \Delta$$

so that $M_{\frac{1}{2}} - M_1$ should be the same for each meson multiplet, independent of spin and parity.

## 57  Mass splitting for baryons

We first attempt to explain the mass splitting within a baryon multiplet by the same procedure used in the previous section for mesons – namely that the mass splitting is entirely due to the mass differences of the quarks. This explanation is not completely successful.

The states of three quarks are shown in Fig. 57.1. If the binding energy $\lambda c^2$ is a constant within a multiplet, the baryon mass is given by

$$M = \sum m_i + \lambda c^2 \qquad (57.1)$$

$\lambda$ will depend on the spin, parity and $SU(3)$ multiplet. $\sum m_i$ is the sum of the masses of the three quarks. This model explains the equal spacing of the masses of the baryon $J^P=\frac{3}{2}^+$ decuplet (see Fig. 49.2, p. 135), but predicts that the mass of the $\Lambda$ and the $\Sigma$ in the octet should be the same. This disagrees with the observed masses for the $J^P=\frac{1}{2}^+$ baryons,

$$M_\Lambda = 1116 \text{ MeV}$$
$$M_\Sigma = 1193 \text{ MeV}$$

A more general symmetry breaking must be considered, and this yields the Gell-Mann–Okubo mass formula, a derivation of which is given in the next section.

We have gone about as far as is possible in the study of particle physics using only extremely simple considerations. To go further requires a certain amount of complication. This is then a convenient point for the reader who has had enough to turn to the closing chapter.

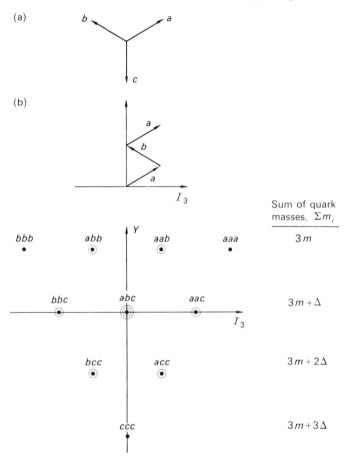

FIGURE 57.1   Construction of baryons from three quarks. (a) The three quarks can be depicted as three vectors in the weight diagram. The baryon states can then be constructed by taking all possible combinations of these three vectors. For instance (b) shows the combination yielding a state with $Y = 1$, $I_3 = +\frac{1}{2}$. (c) All the states obtained by this procedure.

## 58   Derivation of Gell-Mann–Okubo mass formula for octet

For mathematical convenience we consider an octet constructed from a triplet and antitriplet

$$3 \otimes \bar{3} = 8 \oplus 1 \qquad (58.1)$$

The triplet and antitriplet, shown in Fig. 58.1, are not identified with quarks, but are just a convenient mathematical device.

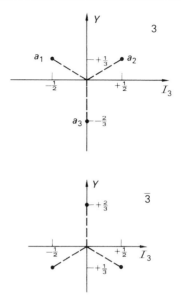

<figure>FIGURE 58.1  Triplet and anti-triplet. Note carefully that $a_1 a_2 a_3$ are not quarks.</figure>

The states of $3 \otimes \bar{3}$ are easily pictured by placing the diagram $\bar{3}$ on each state of the diagram for 3, yielding the states shown in Fig. 58.2.

$a_2$ and $a_1$ constitute an isospin doublet, $I = \frac{1}{2}$, as do also $\bar{a}_2$ and $\bar{a}_1$. $a_2 \bar{a}_1$ and $a_1 \bar{a}_2$ are two states of an isospin triplet, $I = 1$. We use the same symbols for the state function as for the state. Then the third member of the $I = 1$ triplet with $I_3 = 0$ is the symmetric combination of the two $I = \frac{1}{2}$ doublets,

$$\frac{1}{\sqrt{2}} (a_2 \bar{a}_2 + a_1 \bar{a}_1) \tag{58.2}$$

One can introduce a $U$ spin, with the $U_3$ axis at 120° to the $I_3$ axis, as shown in Fig. 58.2, and the states can be arranged in $U$-spin multiplets. Then $a_1$ and $a_3$ constitute a $U$-spin doublet, $U = \frac{1}{2}$, and $a_2$ is a $U$-spin singlet, $U = 0$. The threefold symmetry of the weight diagrams allows the definition of three spins, called $U$ spin, $I$ spin and $V$ spin, with $U_3$, $I_3$ and $V_3$ axes at 120° to each other. However, the use of $V$-spin offers no advantage over the use of only $I$ spin and $U$ spin. Note carefully, that $U$ spin and $V$ spin, like $I$ spin, have nothing whatever to do with angular momentum or spin, but are just a convenient way of handling the mathematics of multiplets. (For a fuller account of the use of $U$ spin, see Lipkin, 1966.)

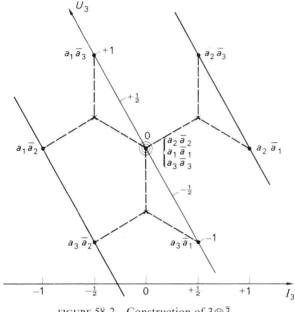

FIGURE 58.2   Construction of $3 \otimes \bar{3}$.

In the same way as the $I = 1$ triplet was considered above by adding the two $I$-spins of $\frac{1}{2}$, the $U$-spin triplet, $U = 1$, obtained by adding the two $U$-spins of $\frac{1}{2}$ is

$$\begin{array}{cc} a_1 \bar{a}_3 & U_3 = +1 \\[6pt] \dfrac{1}{\sqrt{2}}(a_1 \bar{a}_1 + a_3 \bar{a}_3) & U_3 = 0 \\[6pt] a_3 \bar{a}_1 & U_3 = -1 \end{array} \qquad (58.3)$$

This $U$-spin triplet is also part of the $SU(3)$ octet. However, the $U = 1$, $U_3 = 0$ state is not orthogonal to the $I = 1$, $I_3 = 0$ state, equation (58.2), which we choose as the physical state because the observed particles have sharp values of $I$ and $I_3$. As some states of definite $U$-spin are linear combinations of states of different $I$ spin, the observed particles are not always states of good $U$ spin: the observed particles do not always have definite values of $U$, although they do have definite values of $U_3$.

It is convenient to work with orthogonal states, and so we require that part of the $U=1$, $U_3=0$ state,

$$|U = 1, U_3 = 0\rangle$$

which is orthogonal to the $I=1$, $I_3=0$ state

$$|I = 1, I_3 = 0\rangle$$

i.e. we require

$$|U = 1, U_3 = 0\rangle - \langle I = 1, I_3 = 0|U = 1, U_3 = 0\rangle|I = 1, I_3 = 0\rangle \tag{58.4}$$

Now

$$\langle I = 1, I_3 = 0|U = 1, U_3 = 0\rangle = \tfrac{1}{2}\langle(a_2\bar{a}_2+a_1\bar{a}_1)|(a_1\bar{a}_1+a_3\bar{a}_3)\rangle$$
$$= \tfrac{1}{2} \tag{58.5}$$

since

$$\langle a_i\bar{a}_j|a_k\bar{a}_l\rangle = \langle a_i|a_k\rangle\langle\bar{a}_j|\bar{a}_l\rangle$$
$$= \delta_{ik}\delta_{jl} \tag{58.6}$$

since the triplet $a_1a_2a_3$ and antitriplet $\bar{a}_1\bar{a}_2\bar{a}_3$ are independent of each other. $\bar{a}_i$ has nothing to do with $a_i$ here.

The expression (58.4) becomes

$$\frac{1}{\sqrt{2}}(a_1\bar{a}_1+a_3\bar{a}_3) - \frac{1}{2\sqrt{2}}(a_2\bar{a}_2+a_1\bar{a}_1)$$

$$= \frac{1}{2\sqrt{2}}(a_1\bar{a}_1-a_2\bar{a}_2+2a_3\bar{a}_3) \tag{58.7}$$

This state has $I=0$, as both

$$a_3\bar{a}_3$$

and

$$\frac{1}{\sqrt{2}}(a_1\bar{a}_1-a_2\bar{a}_2)$$

have $I=0$, the former because it is a combination of two $I=0$ states, the latter because it is the antisymmetric combination of two $I=\tfrac{1}{2}$ states. The expression (58.7) when normalized is

$$|I = 0\rangle = 6^{-\tfrac{1}{2}}(a_1\bar{a}_1-a_2\bar{a}_2+2a_3\bar{a}_3) \tag{58.8}$$

As there are three states in Fig. 58.2 with $I_3 = 0$, $Y = 0$, there is another $I = 0$ state, but it is an $SU(3)$ singlet and does not interest us for mass splitting.

We can write

$$|U = 1, U_3 = 0\rangle = \langle I = 0|U = 1, U_3 = 0\rangle|I = 0\rangle$$
$$+ \langle I = 1, I_3 = 0|U = 1, U_3 = 0\rangle|I = 1, I_3 = 0\rangle \quad (58.9)$$

Now,

$$\langle I = 0|U = 1, U_3 = 0\rangle$$
$$= 12^{-\frac{1}{2}}\langle(a_1\bar{a}_1 - a_2\bar{a}_2 + 2a_3\bar{a}_3)|(a_1\bar{a}_1 + a_3\bar{a}_3)\rangle$$
$$= \frac{\sqrt{3}}{2} \quad (58.10)$$

Substituting equations (58.10) and (58.5) into (58.9),

$$|U = 1, U_3 = 0\rangle = \frac{\sqrt{3}}{2}|I = 0\rangle + \tfrac{1}{2}|I = 1, I_3 = 0\rangle \quad (58.11)$$

We assume a simple type of symmetry breaking for the mass, namely that within each $U$-spin multiplet the mass is

$$M = M_0 + \alpha U_3 \quad (58.12)$$

Considering the $U$-spin triplet, as shown in Fig. 58.3,

$$\langle U_3 = +1|M|U_3 = +1\rangle = M_0 + \alpha = M_N \quad (58.13)$$

$$\langle U_3 = -1|M|U_3 = -1\rangle = M_0 - \alpha = M_\Xi \quad (58.14)$$

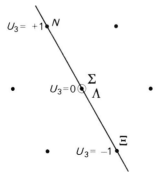

FIGURE 58.3 The $U$-spin triplet in the baryon octet.

$$\langle U_3 = 0 | M | U_3 = 0 \rangle = M_0$$

$$= \left\{ \frac{\sqrt{3}}{2} \langle I = 0 | + \tfrac{1}{2} \langle I = 1, I_3 = 0 | \right\} M \left\{ \frac{\sqrt{3}}{2} | I = 0 \rangle \right.$$
$$\left. + \tfrac{1}{2} | I = 1, I_3 = 0 \rangle \right\}$$

$$= \tfrac{3}{4} \langle I = 0 | M | I = 0 \rangle + \tfrac{1}{4} \langle I = 1, I_3 = 0 | M | I = 1, I_3 = 0 \rangle$$

$$= \tfrac{3}{4} M_\Lambda + \tfrac{1}{4} M_\Sigma \qquad (58.15)$$

From equations (58.13), (58.14) and (58.15),

$$M_N + M_\Xi = \tfrac{1}{2}(3 M_\Lambda + M_\Sigma)$$

which is the Gell-Mann–Okubo mass formula for the baryon octet.

### References

ADAIR, R. K. and H. KASHA, *Phys. Rev. Letters* **23** (1969) 1355.

ANTIPOV, YU. M., N. K. VISHNEVSKIĬ, F. A. ECH, A. M. ZAĬTSEV, I. I. KARPOV, L. G. LANDSBERG, V. G. LAPSHIN, A. A. LEBEDEV, A. G. MOROZOV, YU. D. PROKOSHKIN, YU. V. RODNOV, V. G. RYBAKOV, V. I. RYKALIN, V. A. SEN'KO, B. A. UĬOCHKIN and V. P. KHROMOV, *Sov. J. Nucl. Phys.* **10** (1970a) 199. Translation of *Yadernaya Fizika* **10** (1969) 346.

ANTIPOV, YU. M., V. N. BOLOTOV, N. K. VISHNEVSKIĬ, M. I. DEVISHEV, M. N. DEVISHEVA, F. A. ECH, A. M. ZAĬTSEV, V. V. ISAKOV, I. I. KARPOV, V. A. KRENDELEV, L. G. LANDSBERG, V. G. LAPSHIN, A. A. LEBEDEV, A. G. MOROZOV, YU. D. PROKOSHKIN, V. G. RYBAKOV, V. I. RYKALIN, A. V. SAMOĬLOV, V. A. SEN'KO and YU. S. KHODYREV, *Sov. J. Nucl. Phys.* **10** (1970b) 561. Translation of *Yadernaya Fizika* **10** (1969) 976.

CAIRNS, I., C. B. A. McCUSKER, L. J. PEAK and R. L. S. WOOLCOTT, *Phys. Rev.* **186** (1969) 1394.

CHEW, G. F., M. GELL-MANN and A. H. ROSENFELD, 'Strongly interacting particles', *Sci. Amer.* February 1964. (Also available as reprint 296, Freeman, San Francisco.)

CLARK, A. F., R. D. ERNST, H. F. FINN, G. G. GRIFFEN, N. E. HANSEN, D. E. SMITH and W. M. POWELL, *Phys. Rev. Letters* **27** (1971) 51.

GELL-MANN, M., *Phys. Letters* **8** (1964) 214. Also contained in GELL-MANN and NE'EMAN (1964) and KOKKEDEE (1969).

GELL-MANN, M. and Y. NE'EMAN, *The Eightfold Way*, 1964. Benjamin, New York.

JOYCE, J., *Finnegan's Wake*, 1939. Viking Press. New York. p. 383.

KOKKEDEE, J. J. J., *The Quark Model*, 1969. Benjamin, New York.

LICHTENBERG, D. B., *Unitary Symmetry and Elementary Particles*, 1970. Academic Press, New York.

LIPKIN, H. J., *Lie Groups for Pedestrians*, 2nd edition, 1966. North-Holland, Amsterdam.

MCCUSKER, C. B. A. and I. CAIRNS, *Phys. Rev. Letters* **23** (1969) 658.

MORPURGO, G., *Ann. Rev. Nucl. Sci.* **20** (1970) 105.

SÖDING, P., J. BARTELS, A. BARBARO-GALTIERI, J. E. ENSTROM, T. A. LASINSKI, A. RITTENBERG, A. H. ROSENFELD, T. G. TRIPPE, N. BARASH-SCHMIDT, C. BRICMAN, V. CHALOUPKA, and M. ROOS. *Phys. Letters* **39B** (1972) No. 1.

WEYL, H., *Symmetry*, 1952. Princeton University Press.

ZWEIG, G., 'Symmetries in Elementary Particle Physics', 1965. 1964 International School of Physics *Ettore Majorana* edited by A. Zichichi. Academic Press, New York, p. 192.

### Exercise

1 Using the masses from the latest *Review of Particle Properties* (see end of Section 41), check the accuracy of the Gell-Mann–Okubo mass formula for the baryon octets and decuplets (see Table 51.1).

# Regge poles

# 12

## 59 Regge poles

An important idea in particle physics is that of Regge poles. Regge poles were found by Regge (1959) using the Schrödinger equation to study the analytic properties of the scattered amplitude of particles scattered by a potential. Knowledge of the analytic properties of scattered amplitudes is important in all branches of theoretical physics, but especially in particle physics for which we have no complete dynamical theory.

Consider the scattering of a spinless particle with initial momentum $\hbar k$ and energy

$$E = (\hbar k)^2/2m$$

by a spherically symmetric potential using non-relativistic quantum mechanics. The scattered amplitude can be written as (Saxon, 1968)

$$f(k, \theta) = \sum_{l=0}^{\infty} (2l+1)f_l(k)P_l(\cos \theta) \qquad (59.1)$$

where $f_l(k)$ is the amplitude for the scattering of a particle which has orbital angular momentum $l\hbar$. $\theta$ is the angle of scattering.

For the actual scattering that occurs, $f_l(k)$ is needed only for the physical region, $k$ real and positive. However, a lot can be learned by studying $f_l(k)$ as a function of $k$ for unphysical values of $k$, including complex values of $k$. In the upper half of the complex $k$ plane, $f_l(k)$ is an analytic function of $k$ except for poles and cuts on the imaginary axis.

Regge (1959) showed that it is possible to introduce a function

$$F(l, k)$$

of two complex variables $l$ and $k$, such that $F(l, k)$ coincides with $f_l(k)$ for $l$ a non-negative integer, and such that for Re $l > -\frac{1}{2}$, $F(l, k)$ is an analytic function of $l$ except for poles above or on the real axis. These poles in the complex angular momentum plane are called Regge poles. The positions of the poles are analytic functions of the energy $E$,

$$l = \alpha_i(E), \quad i = 1, 2, 3, \text{ etc.}$$

so that as the energy varies, each Regge pole traces out a trajectory in the complex $l$ plane.

For $E < 0$, there is a bound state at each energy at which $\alpha_i(E)$ passes through a non-negative integer value. For $E > 0$, there are no bound states, but the Regge trajectory may pass close to a non-negative integer for some value of $E$, i.e.

$$\alpha_i(E) = n + i\beta$$

where $n$ is a non-negative integer, and $\beta$ is real and positive. This corresponds to a scattering resonance at the energy $E$ with angular momentum $n\hbar$, while $\beta$ is related to the width of the resonance. The closer the trajectory to the real axis, the narrower the resonance.

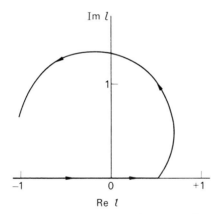

FIGURE 59.1 Regge trajectory. The arrows show the direction in which the Regge pole moves as the energy is increased. An example from Ahmadzadeh *et al.* (1963).

Figure 59.1 shows an example of a Regge trajectory. For $E$ negative, the Regge pole moves along the real $l$ axis, and is at $l = 0$ at the energy of the bound state. At $E = 0$, the Regge pole leaves the real axis and moves into the upper half of the complex $l$ plane. Several examples of Regge trajectories for Yukawa potentials have been calculated by Ahmadzadeh, Burke and Tate (1963), and some of these examples are also given by Omnès and Froissart (1963).

The resonances and bound states can be displayed in families by plotting Re $\alpha$ as a function of $E$. Such plots are shown in Figs. 59.2a and b for Coulomb and harmonic oscillator potentials and in Fig. 59.2c for the energy levels of a rigid rotor.

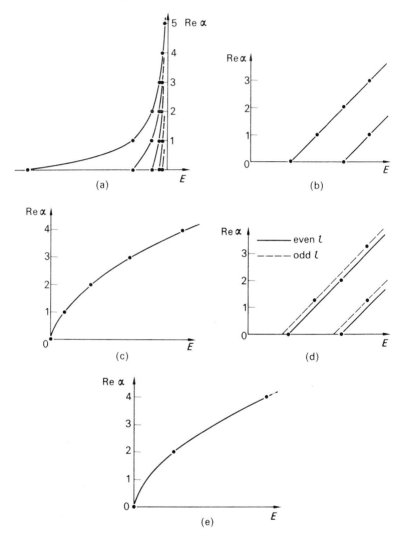

FIGURE 59.2    Regge trajectories. (a) Coulomb potential. (b) Harmonic oscillator potential. (c) Levels of rigid rotor. (d) Effect of exchange forces with harmonic oscillator potential. (e) Rotational levels of a homonuclear diatomic molecule with zero-spin nuclei. The physical states occur at the energies for which Re $\alpha$ is a non-negative even integer.

## 60 Exchange forces

Again using non-relativistic quantum mechanics, consider the scattering of one spinless particle by another when the interaction between the two particles can be described by a potential. Then separating the motion of the centre-of-mass, we are left with the motion in a fixed potential of a particle with the reduced mass, and so the considerations of the previous section can be applied to this problem.

An important factor in the application to particle physics which can be illustrated in non-relativistic theory is that of the effect of exchange forces. Ordinary (non-exchange) forces can be represented in Schrödinger theory by a potential operator $\hat{V}$ such that

$$\hat{V}\psi(\mathbf{r}_1, \mathbf{r}_2) = V(|\mathbf{r}_1 - \mathbf{r}_2|)\psi(\mathbf{r}_1, \mathbf{r}_2)$$

where $V(|\mathbf{r}_1 - \mathbf{r}_2|)$ is the potential. An exchange force is represented by an operator $\hat{V}_e$, which also interchanges the particles,

$$\hat{V}_e\psi(\mathbf{r}_1, \mathbf{r}_2) = V_e(|\mathbf{r}_1 - \mathbf{r}_2|)\psi(\mathbf{r}_2, \mathbf{r}_1)$$

Since

$$\psi(\mathbf{r}_2, \mathbf{r}_1) = (-1)^l\psi(\mathbf{r}_1, \mathbf{r}_2)$$

where $\hbar l$ is the angular momentum about the centre-of-mass, the effect of exchange forces is that there is an effective potential

$$V + V_e$$

for $l$ even and

$$V - V_e$$

for $l$ odd. Thus there will be separate Regge trajectories for $l$ even and for $l$ odd. An example with both $V$ and $V_e$ as harmonic oscillator potentials is shown in Fig. 59.2d. Regge recurrences of each state now occur at intervals of 2 in angular momentum.

A similar phenomenon, but with a different cause (Landau, 1958), occurs in the rotational spectra of homonuclear diatomic molecules (Halliday, 1950; Herzberg, 1939), where there are different sets of rotational states according to whether the spin $J$ is even or odd. For instance, if the nuclei of the molecule have zero spin, they obey Bose statistics and the total wave function must be symmetric with respect to interchange of the two nuclei, and (assuming the usual case of a symmetric electronic wave function) only states with $J$ even can occur, as shown in Fig. 59.2e.

To distinguish between the different types of Regge trajectories, an additional quantum number, known as the signature $\tau$, is introduced. The signature is defined as

$$\tau = (-1)^J$$

for bosons and

$$\tau = (-1)^{J-\frac{1}{2}}$$

for fermions.

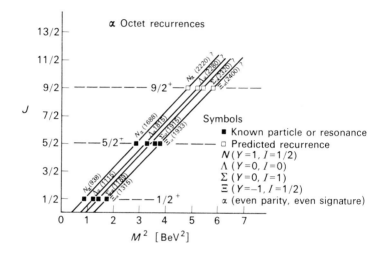

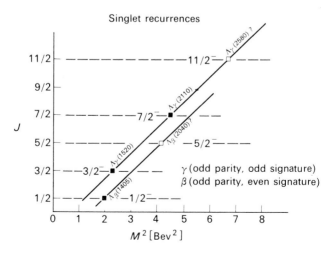

FIGURE 61.1    Chew–Frautschi plots of baryon Regge recurrences classified according to 1, 8 and 10 multiplets of $SU(3)$. (From Barger and Cline, 1967a.)

## 61 Application to particle physics

Even though there is no complete theory of relativistic quantum mechanics, one can speculate that Regge trajectories can be drawn for the strongly interacting particles. The particles lying on a Regge trajectory must all have the same internal quantum numbers such as hypercharge, isospin, etc. The mesons should lie on trajectories with

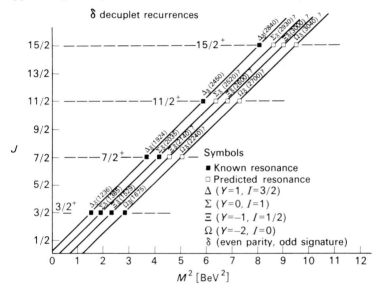

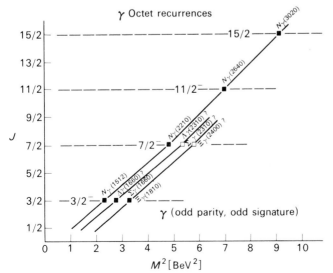

the physically occurring values of $J$ either all even or all odd, so that the Regge recurrences occur at intervals of 2 in $J$.

For the baryons, which have half-odd-integer spins, trajectories are expected with the physical values of $J$ as half-odd-integers again spaced by 2. Taking into account both parity and signature, we expect four kinds of Regge trajectories, which are designated by $\alpha$, $\beta$, $\gamma$, $\delta$ as follows

$\alpha$. $\tau = +1$    $J^P = \frac{1}{2}^+, \frac{5}{2}^+, \frac{9}{2}^+$, etc.

$\beta$. $\tau = +1$    $J^P = \frac{1}{2}^-, \frac{5}{2}^-, \frac{9}{2}^-$, etc.

$\gamma$. $\tau = -1$    $J^P = \frac{3}{2}^-, \frac{7}{2}^-, \frac{11}{2}^-$, etc.

$\delta$. $\tau = -1$    $J^P = \frac{3}{2}^+, \frac{7}{2}^+, \frac{11}{2}^+$, etc.

The most firmly established Regge trajectory of hadrons is that for the $I = \frac{3}{2}$ resonances in meson–nucleon scattering, the $\Delta$'s, for which the spins are plotted against the square of the mass in Fig. 61.1. Such a plot is called a Chew–Frautschi plot (Chew, 1961, 1962). While a smooth curve was expected on a Chew–Frautschi plot, there was no reason to expect such a straight line, and no adequate explanation has been given of why to a remarkably good approximation

$$M^2 \propto^! J$$

As the $\Delta(J^P = \frac{3}{2}^+)$ is part of an $SU(3)$ decuplet, it is expected that the Regge recurrences are also parts of $SU(3)$ decuplets. Of the other members of the $J^P = \frac{3}{2}^+$ decuplet, only a Regge recurrence of the $\Sigma$ has been definitely established. The expected Regge trajectories for the decuplet baryons are shown in Fig. 61.1.

A detailed classification of baryons incorporating Regge poles and $SU(3)$ has been carried out by Barger and Cline (Barger, 1967a, 1967b, 1968), as shown in Fig. 61.1.

The classification by Regge trajectories has not been as useful for mesons as for baryons. The evidence for Regge trajectories for mesons is more indirect, and comes mainly from the analysis of high energy scattering.

The force between two baryons is usually attributed to the exchange of virtual mesons between the two baryons as depicted schematically in Fig. 61.2, where the exchanged meson is taken as a $\pi$-meson. If the exchanged meson lies on a Regge trajectory, then each of the Regge recurrences of the meson can also be exchanged as a virtual particle between the baryons, as shown in Fig. 61.3 for the exchange of a chain of pions with spins 0, 2, 4 and so on. The use of the Regge pole formalism for the scattering amplitude simulates

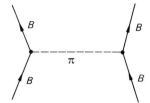

FIGURE 61.2 Schematic diagram depicting the exchange of a π-meson between two baryons.

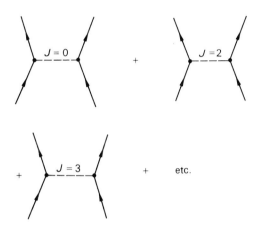

FIGURE 61.3 Schematic diagram depicting the exchange of a π-meson and its Regge recurrences between two baryons. The Regge pole formalism simulates the effect of the exchange of all members of the chain of mesons.

the effect of exchanging in the scattering process all the individual members of the chain of mesons (Barger and Cline, 1969, in particular p. 41). It was hoped that interactions between hadrons at high energies could be described in terms of the exchange of a small number of Regge trajectories, with each Regge trajectory including a known hadron. So far, although there have been some promising indications, this hope has not been realized, and there is some evidence that an alternative description (Leader and Pennington, 1971) is more appropriate for hadronic interactions at high energies.

Evidence indicating that mesons lie on Regge trajectories is also provided by the $I = 1$ mesons as shown in Fig. 45.6 and the result of Exercise 5, Chapter 10.

The baryon and meson Regge trajectories all have much the same slope, namely

$$\frac{\mathrm{d}J}{\mathrm{d}M^2} \simeq 1 \ (\mathrm{GeV})^{-2}$$

## 62 Complications

Although potential scattering in non-relativistic quantum mechanics can be described simply by Regge poles, the situation in relativistic physics is more complicated (Collins, 1968). Mandelstam (1963) showed that there are cuts in the complex $J$ plane as well as poles. It has also been shown that in a relativistic theory, singularities in the complex $J$ plane occur in families. In particular, Regge poles occur in families; each Regge trajectory is accompanied by daughter trajectories. A parent trajectory at $\alpha_0(M^2 = 0)$ implies the existence of daughter trajectories with $\alpha_K(0) = \alpha_0(0) - K$, $K = 1, 2, \ldots$, etc. Some examples of parent and daughter Regge trajectories have been calculated by Chung and Snider (1967).

The theory of Regge poles has had a great impact on the theory of strongly interacting particles, and both its successes and failures have led to many further developments. Unfortunately further treatment of Regge theory would involve a mathematical excursion beyond the scope of this book. The treatment given here has concentrated on the use of Regge trajectories for the classification of hadrons, not only because this aspect of Regge theory is simple but also because this method of classifying hadrons is useful and is capable of predicting the spins and approximate masses of resonances. However, most work on Regge theory is concerned with describing collision processes, and in this respect the Regge pole model is not a theory with a high predictive power.

More detailed information about the application of Regge pole theory to particle classification and to high energy scattering is available in the reviews by Hite (1969) and Barger and Cline (1969). Further promising developments are reviewed by Veneziano (1969).

In conclusion, the theory of Regge poles has already proved useful in high energy physics, even though it is not as useful nor as simple as originally hoped.

### References

AHMADZADEH, A., P. G. BURKE and C. TATE, *Phys. Rev.* **131** (1963) 1315.

BARGER, V. and D. CLINE, *Phys. Rev.* **155** (1967a) 1792.

BARGER, V. and D. CLINE, 'High energy scattering', *Sci. Amer.* December 1967b.

BARGER, V. *Rev. Mod. Phys.* **40** (1968) 129.

BARGER, V. and D. CLINE, *Phenomenological Theories of High Energy Scattering*, 1969. Benjamin, New York.

CHEW, G. F. and S. C. FRAUTSCHI, *Phys. Rev. Letters* **7** (1961) 394.

CHEW, G. F. and S. C. FRAUTSCHI, *Phys. Rev. Letters* **8** (1962) 41.

CHUNG, V. and D. R. SNIDER, *Phys. Rev.* **162** (1967) 1639.

COLLINS, P. D. B. and E. J. SQUIRES, 'Regge poles in particle physics', *Springer Tracts in Modern Physics* **45** (1968).

HALLIDAY, D., *Introductory Nuclear Physics,* 1950. Wiley, New York, p. 484.

HERZBERG, G., *Molecular Spectra and Molecular Structure I*, 1939. Prentice-Hall, New York.

HITE, G. E., *Rev. Mod. Phys.* **41** (1969) 669.

LANDAU, L. D. and E. M. LIFSHITZ, *Quantum Mechanics*, 1958. Pergamon, London, p. 293.

LEADER, E. and M. R. PENNINGTON, *Phys. Rev. Letters* **27** (1971) 1325.

MANDELSTAM, S., *Nuovo Cimento* **30** (1963) 1127:1148.

OMNÈS, R. and M. FROISSART, *Mandelstam Theory and Regge Poles,* 1963. Benjamin, New York.

REGGE, T., *Nuovo Cimento* **14** (1959) 951.

SAXON, D. S., *Elementary Quantum Mechanics*, 1968. Holden-Day, San Francisco.

VENEZIANO, G., *Physics Today* **22** (1969) No. 9, September, p. 31.

### Exercise

1 Using the latest *Review of Particle Properties* by Particle Data Group, prepare an up-to-date version of Fig. 61.1.

# $SU(6)$  # 13

## 63 The quark model and $SU(6)$

The quarks each have spin $\frac{1}{2}$. Including the spin states, there are altogether six quark states

$$\text{two spin states} \times \text{three } SU(3) \text{ states} = \text{six states}$$

Assuming that the interactions of quarks are the same for all of these six states leads to a sixfold symmetry, the group of which is called $SU(6)$. ($SU(6)$ is the group of unitary unimodular $6 \times 6$ matrices.)

The situation of $SU(6)$ in hadron physics is very similar to that of the supermultiplet theory of Wigner in nuclear physics, for which the group is $SU(4)$. The four basic states in supermultiplet nuclear theory are the four states of the nucleon–two spin states each for the proton and the neutron. The assumption of charge independence and spin independence of nuclear forces leads to invariance under $SU(4)$. Since nuclear forces are not spin independent, the $SU(4)$ symmetry is only an approximate symmetry.

Combining a quark and an antiquark, there are $6 \times 6 = 36$ possible meson states, containing the $0^-$ and $1^-$ mesons, which can be split up into two $SU(6)$ multiplets, one a singlet and the other with multiplicity 35. The $SU(6)$ singlet is an $0^-$ meson. The $SU(6)$ multiplet of 35 mesons can be arranged as an $SU(3)$ octet of spin $0^-$ mesons (eight states), an $SU(3)$ octet of spin $1^-$ mesons (24 states, since the spin multiplicity is 3), and an $SU(3)$ singlet of $1^-$ mesons (three states).

Combining three quarks symmetrically with respect to both $SU(3)$ states and spin states, yields an $SU(6)$ multiplet of 56 baryons consisting of an $SU(3)$ octet of spin $\frac{1}{2}$ baryons (16 states), and an $SU(3)$ decuplet of spin $\frac{3}{2}$ baryons (40 states). Thus the low lying baryon states are just those required for an $SU(6)$ multiplet with a multiplicity of 56.

## 64 Ratio of magnetic moments of neutron and proton

One of the most important results of $SU(6)$ applied to particle physics is that taking the neutron and proton as members of the $SU(6)$ 56-plet yields the ratio of their magnetic moments as

$$\mu_p/\mu_n = -\tfrac{3}{2}$$

which is close to the experimental value of $-1\cdot46$. This result is easily obtained using the quark model.

From Section 57, we have that:

> the proton consists of 2 '$a$' quarks and 1 '$b$' quark,
> the neutron consists of 1 '$a$' quark and 2 '$b$' quarks.

For the nucleon, we write the quantum-mechanical operator for the magnetic moment as

$$\hat{\boldsymbol{\mu}} = g\hat{\mathbf{J}} = g_a\hat{\mathbf{J}}_a + g_b\hat{\mathbf{J}}_b \tag{64.1}$$

We use $\hat{\phantom{x}}$ to distinguish operators from their eigenvalues. $\hat{\mathbf{J}}_a$ is the total spin of the one or two $a$ quarks and $\hat{\mathbf{J}}_b$ is the total spin of the two or one $b$ quarks. Then using

$$\hat{\mathbf{J}} = \hat{\mathbf{J}}_a + \hat{\mathbf{J}}_b \tag{64.2}$$

$$
\begin{aligned}
\hat{\mathbf{J}}.\hat{\boldsymbol{\mu}} &= \hat{J}^2 g \\
&= (\hat{\mathbf{J}}_a + \hat{\mathbf{J}}_b).(g_a\hat{\mathbf{J}}_a + g_b\hat{\mathbf{J}}_b) \\
&= g_a\hat{J}_a^2 + g_b\hat{J}_b^2 + (g_a+g_b)\hat{\mathbf{J}}_a.\hat{\mathbf{J}}_b
\end{aligned} \tag{64.3}
$$

Since

$$\hat{\mathbf{J}}_a.\hat{\mathbf{J}}_b = \tfrac{1}{2}(\hat{J}^2 - \hat{J}_a^2 - \hat{J}_b^2)$$

$$\hat{J}^2 g = \tfrac{1}{2}[(g_a+g_b)\hat{J}^2 + (g_a-g_b)(\hat{J}_a^2 - \hat{J}_b^2)] \tag{64.4}$$

Replacing $\hat{J}^2, \hat{J}_a^2, \hat{J}_b^2$ in equation (64.4) by their eigenvalues,

$$J(J+1), J_a(J_a+1), J_b(J_b+1),$$

respectively, we have

$$g = \frac{1}{2}\frac{(g_a+g_b)J(J+1) + (g_a-g_b)\{J_a(J_a+1) - J_b(J_b+1)\}}{J(J+1)} \tag{64.5}$$

The nucleon wave function is totally symmetric, and so must be symmetric with respect to interchange of the spins of the two identical quarks. Thus the two identical quarks in the nucleon have total spin 1.

*Proton*    $J_a = 1, J_b = \frac{1}{2}$

$$g_p = \frac{g_a + g_b}{2} + \frac{5}{3}\frac{g_a - g_b}{2}$$

*Neutron*    $J_a = \frac{1}{2}, J_b = 1$

$$g_n = \frac{g_a + g_b}{2} - \frac{5}{3}\frac{g_a - g_b}{2}$$

Assuming that the $g$ factors of the quarks are proportional to their electric charges

$$g_a = -2g_b$$

and

$$\mu_p/\mu_n = g_p/g_n = -\tfrac{3}{2}$$

For further information about $SU(6)$ and its application to particle physics, see Lichtenberg (1970) or Lipkin (1965, 1966).

### References

LICHTENBERG, D. B., *Unitary Symmetry and Elementary Particles*, 1970. Academic Press, New York.

LIPKIN, H. J., 'Now we are $SU(6)$' in *High-Energy Physics and Elementary Particles–Lectures held at the International Centre for Theoretical Physics, Trieste*, 1965. International Atomic Energy Agency, Vienna.

LIPKIN, H. J., *Lie Groups for Pedestrians*, 2nd edition, 1966. North-Holland, Amsterdam.

# Electromagnetic interactions

# 14

## 65 Introduction

Our knowledge of the electromagnetic interaction is fairly complete. For instance, calculations about purely electromagnetic interactions can be performed to arbitrary accuracy using quantum electrodynamics. Our knowledge of electromagnetic interactions enables us to use them to investigate the properties of elementary particles, and in particular of the strongly interacting particles, the hadrons.

It is very difficult to investigate the properties of hadrons by using the strong interactions, as the strong interactions are so strong that calculations involving them are very difficult and the phenomena due to them are very complicated.

A variety of experiments (Panofsky, 1970; Braben, 1969) have been carried out on electromagnetic interactions of hadrons, for instance, the photoproduction of pions by photons incident on protons and the scattering of photons by protons. However, we will confine our attention to the processes occurring in the scattering of electrons. Processes occurring in electron scattering have an advantage in investigating properties of hadrons over processes occurring by absorption of a photon. In the absorption of a photon of energy $hv$ by a hadron, the momentum transferred to the hadron is $hv/c$, uniquely determined by the energy transfer. In the scattering of an electron, the momentum transfer $\mathbf{p} - \mathbf{p}'$ is not determined by the energy transfer $E - E'$ ($\mathbf{p}$, $E$ are the momentum and energy of the incident electron and $\mathbf{p}'$, $E'$ are the momentum and energy of the scattered electron), i.e. the momentum transfer and energy transfer can be varied independently.

The investigation of the nucleon by the scattering of high energy electrons commenced with the work of Hofstadter at Stanford, and showed that the nucleon is not a point particle but has a finite size. The charge of the proton is distributed over a finite volume.

A convenient anthology of early papers on electron scattering is provided by Hofstadter (1963).

## 66  Form factors

Before dealing with the scattering of electrons by nucleons of finite size, we first consider the much simpler problem of the scattering of spinless particles with charge $e$ by a distribution of electric charge, with total charge $Ze$, spread over a finite region.

Using quantum-mechanical perturbation theory, the matrix element for the scattering of a particle with momentum $\mathbf{p}$ and incident wave function

$$\psi_i = N_i \exp{(i\mathbf{p}.\mathbf{r}/\hbar)} \tag{66.1}$$

to a state with scattered momentum $\mathbf{p}'$ and final wave function

$$\psi_f = N_f \exp{(i\mathbf{p}'.\mathbf{r}/\hbar)} \tag{66.2}$$

can be written

$$\langle \psi_f | V | \psi_i \rangle \tag{66.3}$$

where

$$V(\mathbf{r}) = Ze^2 \int \frac{\rho(\mathbf{r}')}{|\mathbf{r}-\mathbf{r}'|} \, d^3\mathbf{r}' \tag{66.4}$$

where $Ze\rho(\mathbf{r})$ is the charge density. $N_i$, $N_f$ are normalization factors.

$$
\begin{aligned}
\langle \psi_f | V | \psi_i \rangle &= Ze^2 \int \psi_f^*(\mathbf{r}) \int \frac{\rho(\mathbf{r}')}{|\mathbf{r}-\mathbf{r}'|} \, d^3\mathbf{r}' \, \psi_i(\mathbf{r}) \, d^3\mathbf{r} \\
&= N_f^* N_i Ze^2 \int \int e^{i\mathbf{q}\cdot\mathbf{r}} \frac{\rho(\mathbf{r}')}{|\mathbf{r}-\mathbf{r}'|} \, d^3\mathbf{r}' \, d^3\mathbf{r}
\end{aligned} \tag{66.5}
$$

where $\mathbf{q} = (\mathbf{p}-\mathbf{p}')/\hbar$ is the momentum transfer in units of $\hbar$. Changing variables to $\boldsymbol{\xi} = \mathbf{r}-\mathbf{r}'$ and $\mathbf{r}'$,

$$
\begin{aligned}
\langle \psi_f | V | \psi_i \rangle &= N_f^* N_i Ze^2 \int \frac{e^{i\mathbf{q}\cdot\boldsymbol{\xi}}}{\xi} \, d^3\boldsymbol{\xi} \int e^{i\mathbf{q}\cdot\mathbf{r}'} \rho(\mathbf{r}') \, d^3\mathbf{r}' \\
&= X(q)F(\mathbf{q})
\end{aligned} \tag{66.6}
$$

where

$$X(q) = N_f^* N_i Ze^2 \int \frac{e^{i\mathbf{q}\cdot\mathbf{r}}}{r} \, d^3\mathbf{r} \tag{66.7}$$

is the matrix element for the scattering by a point charge, and

$$F(\mathbf{q}) = \int e^{i\mathbf{q}\cdot\mathbf{r}} \rho(\mathbf{r}) \, d^3\mathbf{r} \tag{66.8}$$

is called the form factor. The form factor is the Fourier transform of the charge distribution.

The scattering cross-section is proportional to

$$|\langle \psi_f | V | \psi_i \rangle|^2$$

and so is given by

$$\frac{d\sigma}{d\Omega} = \left(\frac{d\sigma}{d\Omega}\right)_{\text{point}} |F(\mathbf{q})|^2 \tag{66.9}$$

$(d\sigma/d\Omega)_{\text{point}}$ is the differential cross-section for the scattering by a point charge, which is called Rutherford scattering.

$$\left(\frac{d\sigma}{d\Omega}\right)_{\text{point}} = \left(\frac{Ze^2 E}{2c^2 p^2}\right)^2 \frac{1}{\sin^4 \frac{1}{2}\theta} \tag{66.10}$$

where $Ze$ is the charge of the scatterer. (The non-relativistic Rutherford formula is obtained by replacing $E$ by $mc^2$ in equation (66.10).)

We restrict our attention to spherically symmetric charge distributions, i.e. $\rho(\mathbf{r}) = \rho(r)$. For small $q$, equation (66.8) yields

$$\begin{aligned}
F(q) &= \int \left(1 + i\mathbf{q}.\mathbf{r} - \frac{(\mathbf{q}.\mathbf{r})^2}{2} + \cdots\right)\rho(r)\, d^3r \\
&= \int \rho(r)\, d^3\mathbf{r} - \frac{2\pi}{3} q^2 \int \rho(r) r^4\, dr + \cdots \\
&= 1 - \frac{q^2}{6} \int \rho(r) r^2\, d^3\mathbf{r} + \cdots \\
&= 1 - \frac{q^2}{6} \langle r^2 \rangle + \cdots
\end{aligned} \tag{66.11}$$

where $\langle r^2 \rangle$ is the mean square radius of the charge distribution. At a particular energy, the maximum value of $q$ is $2p\hbar^{-1}$. At sufficiently low energies such that

$$p^2\hbar^{-2}\langle r^2 \rangle \ll 1$$

we have

$$F(q) \simeq 1$$

and the scattering is the same as the scattering by a point charge.

For extremely high energy particles such that $E \gg mc^2$, equation (66.10) becomes

$$\left(\frac{d\sigma}{d\Omega}\right)_{\text{point}} = \left(\frac{Ze^2}{2E}\right)^2 \frac{1}{\sin^4 \frac{1}{2}\theta} \tag{66.12}$$

The scattering of energetic electrons by a point charge $e$, including the effect of the spin of the electron, is called Mott scattering and is given by

$$\left(\frac{d\sigma}{d\Omega}\right)_{\text{Mott}} = \left(\frac{e^2}{2E}\right)^2 \frac{\cos^2 \frac{1}{2}\theta}{\sin^4 \frac{1}{2}\theta} \qquad (66.13)$$

If the target point charge has mass $M$ but no spin, the scattering cross-section in the laboratory frame is given by

$$\left(\frac{d\sigma}{d\Omega}\right)_{\text{NS}} = \left(\frac{e^2}{2E}\right)^2 \frac{\cos^2 \frac{1}{2}\theta}{\sin^4 \frac{1}{2}\theta} \frac{1}{1+(2E/Mc^2)\sin^2 \frac{1}{2}\theta} \qquad (66.14)$$

The last term in equation (66.14) is due to the recoil of the target particle.

## 67 The form factors of the proton

The theory of the scattering of electrons by protons is considerably more complicated than the simple theory of the previous section, as the proton has spin and so a magnetic moment which also scatters the electron. Thus the proton has two form factors, one to describe the effects of the finite extension in space of its charge, and the other to describe the effects of the finite extension in space of its magnetic moment.

To ensure relativistic invariance of the theory, the form factors of the proton are expressed as functions of the square of the invariant four-momentum transfer (see Appendix A, Section A2),

$$(\mathbf{p}-\mathbf{p}')^2 - c^{-2}(E-E')^2$$

For the remainder of this chapter, we shall use natural units, $\hbar = c = 1$. Then the square of the invariant four-momentum transfer can be written as

$$Q^2 = (\mathbf{p}-\mathbf{p}')^2 - (E-E')^2 \qquad (67.1)$$

and is given by

$$Q^2 = \frac{4E^2 \sin^2 \frac{1}{2}\theta}{1+(2E/M)\sin^2 \frac{1}{2}\theta} \qquad (67.2)$$

when $E$ is the energy of the incident electron in the laboratory frame, in which the proton is initially at rest. $M$ is the mass of the proton.

The scattering of an electron by a proton can be considered as an exchange of a virtual photon between the electron and proton, as shown in Fig. 67.1, and the cross-section is given by the Rosenbluth formula,

$$\frac{d\sigma}{d\Omega} = \left(\frac{d\sigma}{d\Omega}\right)_{NS} \left\{ \frac{G_E^2(Q^2) + (Q^2/4M^2)G_M^2(Q^2)}{1 + Q^2/4M^2} \right.$$

$$\left. + 2 \tan^2 \tfrac{1}{2}\theta \, \frac{Q^2}{4M^2} \, G_M^2(Q^2) \right\} \quad (67.3)$$

$G_E(Q^2)$ is the charge form factor or electric form factor of the proton, and $G_M(Q^2)$ is the magnetic form factor of the proton.

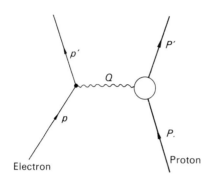

FIGURE 67.1    Elastic electron–proton scattering. The blob at the photon–proton vertex signifies that the proton is not a point particle, and is described by form factors.

At small energies, the scattering from the proton is the same as the scattering from a point charge and a point magnetic moment, so that

$$G_E(0) = 1 \qquad\qquad (67.4)$$

$$G_M(0) = \mu_p = 2\cdot793 \qquad\qquad (67.5)$$

where $\mu_p$ is the magnetic moment of the proton expressed in nuclear magnetons.

Because of the effect of the recoil of the proton, the form factors of the proton cannot be simply related by Fourier transforms to the charge distribution or distribution of magnetic moment in a way analogous to equation (66.8). A more detailed treatment of proton form factors is given by Griffy and Schiff (1967).

Some of the experimental results for the form factors of the proton are shown in Figs. 67.2 and 67.3. To a good approximation, the

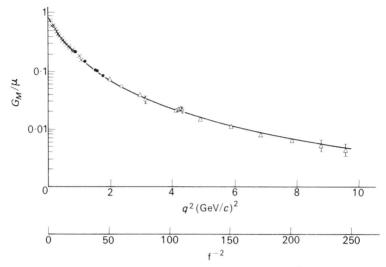

FIGURE 67.2    The magnetic form factor of the proton. $q^2$ of the figure is $Q^2$ of the text. The curve is the dipole fit, equation (67.7). (From Albrecht *et al.*, 1967.)

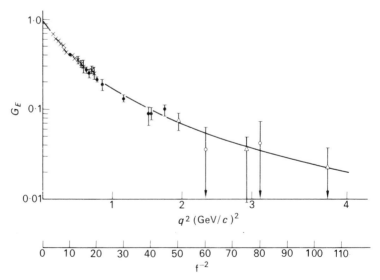

FIGURE 67.3    The electric form factor of the proton. $q^2$ of the figure is $Q^2$ of the text. The curve is the electric form factor given by the dipole fit and the scaling law. (From Albrecht *et al.*, 1967.)

experimental results are consistent with

$$G_E(Q^2) = G_M(Q^2)/\mu_p \qquad (67.6)$$

which is called the scaling law, and with the so-called dipole fit

$$G_M(Q^2)/\mu_p = G_D(Q^2) \qquad (67.7)$$

where

$$G_D(Q^2) \equiv [1 + Q^2/D]^{-2} \qquad (67.8)$$

with $D = 0.71$ (GeV/$c$)$^2$. In Fig. 67.2, the dipole fit, equation (67.7), is compared with some of the experimental results. The electric form factor given by equations (67.7) and (67.6) is shown in Fig. 67.3.

It is still not clear experimentally whether there are any deviations from the scaling law, equation (67.6) (Rutherglen, 1969). However, it is clear that the experimental results show small but systematic deviations from the dipole fit. For example, the data of Bartel *et al.* (1970) and Berger *et al.* (1971) for $(G_M/\mu)/G_D$ are shown in Fig. 67.4.

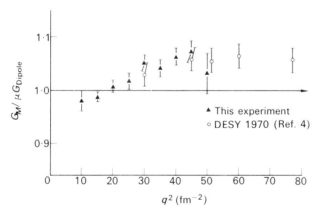

FIGURE 67.4    The magnetic form factor $G_M$, normalized to the dipole fit. $q^2$ of the figure is $Q^2$ of the text. Ref. 4 of the figure is Bartel *et al.* (1970). (From Berger *et al.*, 1971.)

Since we are using natural units with $\hbar = 1$, $Q^2$ can be the square of momentum transfer, or alternatively can be the square of momentum transfer divided by $\hbar^2$ and have the dimensions of (length)$^{-2}$. $Q^2$ is usually expressed in units of (GeV/$c$)$^2$ or in units of fm$^{-2}$. 'fm' or 'f' is used as an abbreviation for 'fermi'.

$$1 \text{ fm} = 10^{-13} \text{ cm}$$

## 68  The form factors of the neutron

The neutron can scatter electrons because of its magnetic moment. Also, although the total charge of the neutron is zero, it may have a non-zero charge density. The scattering of electrons by the neutron is also given by the Rosenbluth formula, equation (67.3), but as the total charge of the neutron is zero,

$$G_E^{\text{neutron}}(0) = 0 \tag{68.1}$$

$$G_M^{\text{neutron}}(0) = \mu_n = -1 \cdot 913 \tag{68.2}$$

where $\mu_n$ is the magnetic moment of the neutron in nuclear magnetons.

The scattering of electrons by the neutron cannot be investigated directly by experiment, but the cross-sections for the scattering of electrons by neutrons can be determined from experiments on the scattering of electrons by deuterons. Because of this, the measurements of neutron form factors are less accurate than proton form factors. The experimental results have been reviewed by Rutherglen (1969). The data on $G_M^{\text{neutron}}$ are not in disagreement with the dipole fit,

$$G_M^{\text{neutron}}(Q^2)/\mu_n = G_D(Q^2) \tag{68.3}$$

with $G_D$ given by equation (67.8), but the precision of the experimental data is not good enough to test whether there are deviations from the dipole fit of the same magnitude as for the proton. So that, within experimental error,

$$G_M^{\text{neutron}}(Q^2)/\mu_n = G_M^{\text{proton}}(Q^2)/\mu_p \tag{68.4}$$

The charge form factor for the neutron for small $Q^2$ can be investigated by the scattering of thermal neutrons by atomic electrons, yielding the result

$$\left(\frac{dG_E^{\text{neutron}}}{dQ^2}\right)_{Q^2=0} = (0 \cdot 50 \pm 0 \cdot 01)\left(\frac{\text{GeV}}{c}\right)^{-2} \tag{68.5}$$

The results of electron scattering experiments are consistent with equation (68.5) for small $Q^2$, and otherwise show only that $G_E^{\text{neutron}}$ is small.

There is as yet no satisfactory theory of the electromagnetic form factors of the nucleon. A comparison of theory and experiment for the form factors is given by Rutherglen (1969). An interesting feature of the theory of nucleon form factors, is that earlier theory (Frazer and Fulco, 1959) predicted the existence of meson resonances, which were subsequently discovered in other experiments. On the other hand, the theory failed to give nucleon form factors in quantitative agreement with experiment. An account of earlier theory of nucleon form factors is given by Muirhead (1965).

## 69 Inelastic scattering

At high energies, as well as elastic scattering, the production of one or more pions may occur in the scattering of electrons by nucleons,

$$e^- + N \rightarrow e^- + N + \pi$$
$$e^- + N \rightarrow e^- + N + \pi + \pi$$

and at sufficiently high energies other particles may also be produced.

Figure 69.1 shows some of the experimental results of Bartel *et al.* (1968) on the inelastic scattering of electrons by protons. The peaks in the spectrum of scattered electrons correspond to excitation of excited states of the nucleon which then decay by emission of pions, i.e.

$$e^- + N \rightarrow e^- + N^*$$

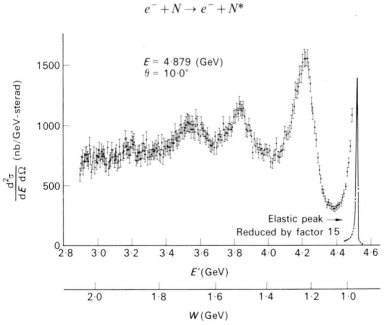

FIGURE 69.1 An example of an electron spectrum for inelastic *e–p* scattering, before application of radiative corrections. (From Bartel *et al.*, 1968.)

It may be noted in Fig. 69.1 that the elastic peak at the right is asymmetric with a larger tail to the left. This is because in all collisions of electrons, some photons are emitted, i.e. the electron always loses some energy by electromagnetic radiation. The theoretical expressions, such as equation (67.3) for elastic scattering, and equation (69.1) which follows for inelastic scattering, apply to the cross-sections after radiative corrections have been made to allow for the effect of the emission of photons.

To the left of the peaks in Fig. 69.1, there is a smoother continuous spectrum of scattered electrons. As the energy of the incident electrons is increased, the peaks tend to disappear but the continuous spectrum persists.

An interesting feature of the experimental results is the weak dependence on momentum transfer of the inelastic cross-sections for excitations well beyond the resonance region. Figure 69.2 shows a demonstration by Briedenbach *et al.* (1969) of this weak dependence. The differential cross-section divided by the Mott cross-section, $(\mathrm{d}^2\sigma/\mathrm{d}\Omega\,\mathrm{d}E')/(\mathrm{d}\sigma/\mathrm{d}\Omega)_{\mathrm{Mott}}$, is plotted as a function of the square of the four-momentum transfer, $Q^2 = 2EE'(1-\cos\theta) = q^2$, for constant values of the invariant mass of the recoiling target system, $W$, where $W^2 = 2M(E-E') + M^2 - Q^2$. The cross-section is divided by the Mott cross-section, equation (66.13), to remove the major part of the four-momentum transfer dependence. The results in Fig. 69.2 for each value of $W$ are from measurements at both $\theta = 6°$ and $10°$. The elastic cross-section, divided by the Mott cross-section for $\theta = 10°$, is included in Fig. 69.2, in order to demonstrate the striking difference between the behaviour of the inelastic and elastic cross-sections. The relatively slow variation with $Q^2$ of the inelastic cross-section compared with the elastic cross-section is clearly shown in Fig. 69.2. For elastic scattering, $\sigma/\sigma_{\mathrm{Mott}}$ falls off rapidly with increasing $Q^2$ because of the finite size of the proton. For inelastic scattering, the weak dependence of $\sigma/\sigma_{\mathrm{Mott}}$ on $Q^2$ can be explained by assuming that the electrons are scattered by constituents of the proton, and that these constituent parts of the proton are very much smaller in size than the proton. Feynman (1969) introduced the name 'partons' for these hypothetical constituent parts of the proton, and assumed that they are point charges.

The process of inelastic scattering of an electron by a nucleon is shown schematically in Fig. 69.3. An electron of energy $E$ scatters off a nucleon (of four-momentum $P$) at an angle $\theta$ and with final state $n$ by means of an exchange of a virtual photon (of four-momentum $Q$). We consider experiments in which only the scattered

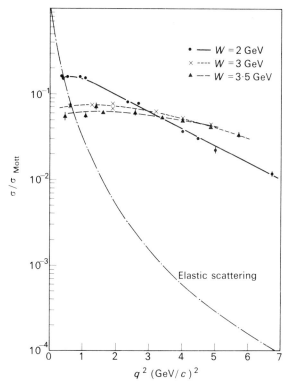

FIGURE 69.2   $(d^2\sigma/d\Omega\,dE')/\sigma_{\text{mott}}$ in $\text{GeV}^{-1}$, vs $q^2$ ($\equiv Q^2$ of text) for $W = 2, 3$ and $3\cdot 5$ GeV. The lines drawn through the data are meant to guide the eye. Also shown is the cross-section for elastic $e$–$p$ scattering divided by $\sigma_{\text{Mott}}$, $(d\sigma/d\Omega)/\sigma_{\text{Mott}}$, calculated for $\theta = 10^\circ$, using the dipole form factor, equation (67.7). The relatively slow variation with $Q^2$ of the inelastic cross-section compared with the elastic cross-section is clearly shown. (From Briedenbach *et al.*, 1969.)

electrons are observed, i.e. the details of the final hadron state $n$ are not observed. Then the most general form for scattering cross-section in the laboratory frame is given by

$$\frac{d^2\sigma}{d\Omega\,dE'} = \frac{4e^4}{Q^4}\,E'^2\cos^2\tfrac{1}{2}\theta[W_2 + 2W_1\tan^2\tfrac{1}{2}\theta]$$
$$= \frac{e^4}{4E^2\sin^4\tfrac{1}{2}\theta}\cos^2\tfrac{1}{2}\theta[W_2 + 2W_1\tan^2\tfrac{1}{2}\theta] \qquad (69.1)$$

where $W_1$ and $W_2$ depend on the hadron physics. As details of the final hadron state are not observed, $W_1$ and $W_2$ must be functions of the relativistically invariant variables, $Q^2$ and

$$v = -\frac{1}{M}Q.P \tag{69.2}$$

$Q.P$ is the invariant product of two 4-vectors, see Appendix A, Section A2. The factor $1/M$ is used for convenience to agree with general notation.

$$v = -\frac{1}{M}\{\mathbf{q}.\mathbf{P}-(E-E')(\mathbf{P}^2+M^2)^{\frac{1}{2}}\}$$

In the laboratory frame, $\mathbf{P}=0$ and

$$v = E-E' \tag{69.3}$$

$$Q^2 = 4EE'\sin^2\tfrac{1}{2}\theta \tag{69.4}$$

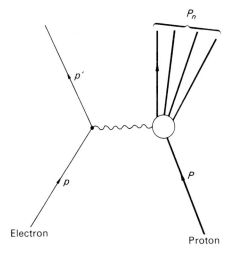

FIGURE    69.3    Inelastic    electron-nucleon scattering.

We now consider the parton model of the nucleon, following the treatment of Bjorken and Paschos (1969). In the parton model, the inelastic scattering of electrons by nucleons is considered to be elastic scattering by partons. To deal with this idea further, we first consider the elastic scattering from a free point particle of mass $m$, with initial momentum and energy $\mathbf{P}_m$, $E_m$, and final momentum and energy $\mathbf{P}'_m$, $E'_m$. We examine the scattering in the frame of reference in which $P_m = 0$. Then, $E_m = m$, and $P_m'^2 = q^2$ and

$$-Q.P_m = (E-E')m = (E_m' - m)m$$
$$Q^2 = q^2 - (E-E')^2 = q^2 - (E_m' - m)^2$$
$$= q^2 + m^2 - E_m'^2 + 2(E_m' - m)m$$

But

$$E_m'^2 = q^2 + m^2$$

so that

$$Q^2 = -2Q.P_m \tag{69.5}$$

in this frame. But both sides of equation (69.5) are relativistically invariant, and so the equation holds in an arbitrary reference frame. For the scattering by this particle, then

$$W_2(v, Q^2) = K \delta(v - Q^2/2m) \tag{69.6}$$

where the constant of proportionality $K$ can be determined by requiring that substituting (69.6) into (69.1) yields point charge scattering. $\delta(y)$ is the Dirac delta function defined by

$$\delta(y) = 0 \quad \text{for } y \neq 0$$

and

$$\int f(y)\delta(y) = f(0)$$

(Feynman, 1965; Merzbacher, 1970). We have

$$\frac{d\sigma}{d\Omega} = \int dE' \frac{d\sigma}{d\Omega\, dE'}$$
$$= K \int dE' \frac{e^4}{4E^2 \sin^4 \frac{1}{2}\theta} \cos^2 \frac{1}{2}\theta\; \delta\left(E - E' - \frac{2EE'}{m} \sin^2 \frac{1}{2}\theta\right)$$
$$= K \frac{e^4}{4E^2 \sin^4 \frac{1}{2}\theta} \frac{\cos^2 \frac{1}{2}\theta}{[1 + (2E/m) \sin^2 \frac{1}{2}\theta]}$$
$$= \left(\frac{d\sigma}{d\Omega}\right)_{NS}$$

as required, if $K = 1$. The property

$$\delta(ay) = |a|^{-1} \delta(y) \tag{69.7}$$

of the $\delta$ function has been used. Equation (69.6) can be written

$$W_2 = \delta\left(\frac{P_m.Q}{m} - \frac{Q^2}{2m}\right) \tag{69.8}$$

If the inelastic scattering of an electron by a proton is viewed in a frame of reference in which the proton has infinite momentum, the motion of the constituent partons within the proton is slowed down by relativistic time dilatation, and the electron can be viewed as scattering instantaneously off an individual parton. The parton recoils and will interact with the rest of the proton to produce the final hadron state, but this interaction with the rest of the proton occurs later, and the parton can be considered to first recoil as a free particle. The inelastic scattering of an electron by a proton according to the parton model is shown schematically in Fig. 69.4. The proton is initially considered to be composed of an arbitrary number of partons, one of which scatters the electron. The momentum–energy four-vector of the parton is taken as a fraction $x$ of the momentum–energy four-vector of the proton,

$$P_m = xP \tag{69.9}$$

in a frame of reference in which the proton has infinite momentum. The electron-proton centre-of-mass frame, at high energies, is a good approximation to such a frame. Then the contribution to

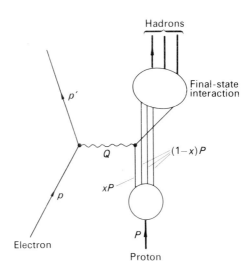

FIGURE 69.4   Inelastic electron–nucleon scattering according to the parton model. The nucleon can be considered as composed of partons, one of which scatters the electron and recoiling interacts with the other partons to produce the final state.

$W_2$ from a single parton of charge $Z_i$ is obtained from equation (69.8),

$$W_2^{(i)} = Z_i^2 \, \delta\left(\frac{xP.Q}{m} - \frac{Q^2}{2m}\right)$$

$$= Z_i^2 \, \delta\left(\frac{P.Q}{M} - \frac{Q^2}{2Mx}\right)$$

$$= Z_i^2 \, \delta\left(v - \frac{Q^2}{2Mx}\right) \tag{69.10}$$

since from equation (69.9), the effective rest mass of the parton is given by

$$m = xM \tag{69.11}$$

For a general distribution of partons in the proton, we have

$$W_2(v, Q^2) = \sum_N P(N) \left\langle \sum_i Z_i^2 \right\rangle_N \int_0^1 dx \, f_N(x) \, \delta\left(v - \frac{Q^2}{2xM}\right) \tag{69.12}$$

where $P(N)$ is the probability of finding a configuration of $N$ partons in the proton, $\langle \sum_i Z_i^2 \rangle_N$ equals the average value of $\sum_i Z_i^2$ in such configurations, and $f_N(x)$ is the probability of finding in such configurations a parton with fraction $x$ of the proton's momentum.

Changing the variable of integration to $Q^2/2xM$, and integrating, equation (69.12) yields

$$vW_2(v, Q^2) = \sum_N P(N) \left\langle \sum_i Z_i^2 \right\rangle_N xf_N(x)$$

$$\equiv F(x) \tag{69.13}$$

with

$$x = Q^2/2Mv \tag{69.14}$$

Therefore $vW_2$ is predicted to be a function of a single variable $x$.

Figure 69.5 shows that this prediction of the parton model agrees with the experimental results, providing evidence that protons are composed of point particles.

It is possible, but not necessarily so, that partons may be identical with quarks. However, it can be shown (Bjorken and Paschos, 1969) that a model in which the proton is composed of three quarks is in disagreement with the experimental results in inelastic electron–proton scattering. The quark model and parton model are consistent if the proton is composed of three quarks and an indefinite number of quark-antiquark pairs.

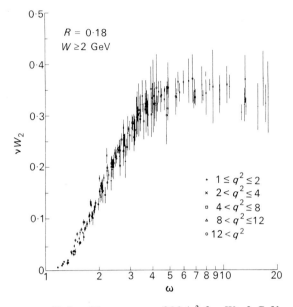

FIGURE 69.5 $\nu W_2$ versus $\omega = 2M\nu/q^2$ for $W > 2$ GeV and $q^2 > 1$ (GeV/$c$)$^2$. The ranges of $q^2$ are given in (GeV/$c$)$^2$. Data from each angle are shown. Note that $q^2$ is $Q^2$ of the text. (From Bloom, E. D., G. Buschhorn, R. L. Cottrell, D. H. Coward, H. De Staebler, J. Drees, C. L. Jordan, G. Miller, L. Mo, H. Piel, R. E. Taylor, M. Briedenbach, W. R. Ditzler, J. I. Friedman, G. C. Hartmann, H. W. Kendall and J. S. Poucher, *Recent Results in Inelastic Electron Scattering*. A report presented to the XV International Conference on High Energy Physics, Kiev, U.S.S.R. SLAC-PUB-796, September 1970.)

It is not pretended that the treatment of the parton model given here is logical, self-consistent or even sensible. The important feature of the parton model is that it agrees with certain experimental results. The possibility cannot be excluded that some more mundane explanation will be found to replace the parton model. Another possibility is that future development of the parton model and quark model may lead eventually to a more satisfactory theory, as in an analogous way Bohr's theory of quantized orbits was a fore-runner of quantum mechanics.

Further information on partons is given by Drell (1970). An interesting account of the structure of the nucleon, as shown by elastic and inelastic scattering of electrons, is given by Kendall and Panofsky (1971).

## 70 $e^+ - e^-$ colliding beams

Experiments with $e^+$–$e^-$ colliding beams are providing important results. Such experiments have been performed at Orsay, Novosibirsk and Frascati. Among the reactions which have been studied are

$$e^+ + e^- \rightarrow e^+ + e^- \quad \text{(elastic scattering)} \quad (70.1)$$
$$\rightarrow \mu^+ + \mu^- \quad (70.2)$$
$$\rightarrow \pi^+ + \pi^- \quad (70.3)$$
$$\rightarrow \pi^+ + \pi^- + \pi^0 \quad (70.4)$$
$$\rightarrow K^+ + K^- \quad (70.5)$$

The properties of boson resonances can be investigated by $e^+$–$e^-$ colliding beam experiments, since for instance, reactions (70.3) and (70.5) proceed to a large extent by the production of resonances

$$e^+ + e^- \rightarrow \rho^0 \rightarrow \pi^+ + \pi^- \quad (70.6)$$

$$e^+ + e^- \rightarrow \phi \rightarrow K^+ + K^- \quad (70.7)$$

Further information on $e^+$–$e^-$ experiments is given by Buon (1970), Panofsky (1970), Pellegrini and Sessler (1970), Sidorov (1969).

### References

ALBRECHT, W., H.-J. BEHREND, H. DORNER, W. FLAUGER and H. HULTSCHIG, *Phys. Rev. Letters* **18** (1967) 1014.

BARTEL, W., B. DUDELZAK, H. KREHBIEL, J. MCELROY, U. MEYER-BARKHOUT, W. SCHMIDT, V. WALTHER and G. WEBER, *Phys. Letters* **28**B (1968) 148.

BARTEL, W., F. W. BÜSZER, W. R. DIX, R. FELST, D. HARMS, H. KREHBIEL, P. E. KUHLMAN, J. MCELROY and G. WEBER, *Phys. Letters* **33**B (1970) 245.

BERGER, CH., V. BURKERT, G. KNOP, B. LANGENBECK and K. RITH, *Phys. Letters* **35**B (1971) 87.

BJORKEN, J. D. and E. A. PASCHOS, *Phys. Rev.* **185** (1969) 1975.

BRABEN, D. W., editor, *Proceedings 4th International Symposium on Electron and Photon Interactions at High Energies, Liverpool.* Daresbury Nuclear Physics Laboratory, 1969.

BRIEDENBACH, M., J. I. FRIEDMAN, H. W. KENDALL, E. D. BLOOM, D. H. COWARD, H. DE STAEBLER, J. DREES, L. W. MO and R. E. TAYLOR, *Phys. Rev. Letters* **23** (1969) 935.

BUON, J., *Particles and Nuclei* **1** (1970) 141.

DRELL, S. D., *Comments on Nuclear and Particle Physics* **4** (1970) 147.

FEYNMAN, R. P., R. B. LEIGHTON and M. SANDS, *Quantum Mechanics,* Vol. III of *The Feynman Lectures in Physics,* 1965. Addison-Wesley, Reading, Mass. Section 16.4.

FEYNMAN, R. P., 'The behaviour of Hadron collisions at extreme energies', *High Energy Physics – Third International Conference,* p. 237. Stony Brook, 1969. Gordon and Breach, New York.

FRAZER, W. R. and J. R. FULCO, *Phys. Rev. Letters* **2** (1959) 365.

GRIFFY, T. A. and L. I. SCHIFF, 'Electromagnetic Form Factors', p. 341 of *High Energy Physics,* Vol. I, edited E. H. S. Burhop, 1967. Academic Press, New York.

HOFSTADTER, R., *Electron Scattering and Nuclear and Nucleon Structure,* 1963. Benjamin, New York.

KENDALL, H. W. and W. K. H. PANOFSKY, 'The structure of the proton and the neutron', *Sci. Amer.,* June 1971, p. 61.

KROHN, V. E. and G. R. RINGO, *Phys. Rev.* **148** (1966) 1303.

MERZBACHER, E., *Quantum Mechanics,* 2nd edition., 1970. Wiley, New York.

MUIRHEAD, H., *The Physics of Elementary Particles,* 1965. Pergamon Press, Oxford. Chapter 11.

PANOFSKY, W. K. H., *Comments on Nuclear and Particle Physics* **4** (1970) 159.

PELLEGRINI, C. and A. M. SESSLER, *Comments on Nuclear and Particle Physics* **4** (1970) 55.

RUTHERGLEN, J. G., 'Nucleon form factors', p. 163 of Braben (1969).

SIDOROV, V. A., 'Storage rings – Novosibirsk', p. 227 of Braben (1969).

# Epilogue

15

Our knowledge of the physics of elementary particles is always changing. No doubt at least some of the material included in this book will eventually turn out to be incorrect. The reader who wishes to learn about later developments is referred to the various reviews which appear in *Comments on Nuclear and Particle Physics*, *Reviews of Modern Physics*, *Annual Review of Nuclear Science* and *Physics Today*. *Scientific American* is especially useful, as an idea or result in high energy physics has not been thoroughly digested and understood until it appears there, and maybe not even then.

This book is intended to provide a knowledge of particle physics suitable for the non-specialist. The reader who wishes to gain a more advanced knowledge should proceed to some of the references listed below.

BURHOP, E. H. S., editor, *High Energy Physics*, 1967. Academic Press, New York. Three volumes.

FELD, B. T., *Models of Elementary Particles*, 1969. Blaisdell, Waltham, Mass.

GASIOROWICZ, S., *Elementary Particle Physics*, 1966. Wiley, New York.

KÄLLÉN, G., *Elementary Particle Physics*, 1964. Addison-Wesley, Reading, Mass.

LICHTENBERG, D. B., *Unitary Symmetry and Elementary Particles*, 1970. Academic Press, New York.

MUIRHEAD, H., *The Physics of Elementary Particles*, 1965. Pergamon, Oxford.

PILKUHN, H., *The Interactions of Hadrons*, 1967. North-Holland, Amsterdam.

# Summary of special relativity[†]

## A1 Introduction

There exist frames of reference in which a free particle remains at rest or moves with constant velocity.

The principle of special relativity states that all inertial frames of reference are equivalent for the formulation of physical laws. In particular, the velocity of light is the same in all inertial frames.

The transformation of coordinates from one inertial frame $F$ with coordinates $x$, $y$, $z$, $t$ to another inertial frame $F'$ is given by a Lorentz transformation. For the case where $x$, $y$, $z$ axes are parallel to $x'y'z'$ axes respectively, the origins of $F$ and $F'$ coincide at $t = t' = 0$, and $F'$ moves in the direction of the positive $x$ axis of $F$ with constant velocity $v$, the Lorentz transformation is

$$x' = \gamma(x - vt) \qquad (A.1)$$

$$y' = y \qquad (A.2)$$

$$z' = z \qquad (A.3)$$

$$t' = \gamma(t - vx/c^2) \qquad (A.4)$$

where

$$\gamma = (1 - \beta^2)^{-\frac{1}{2}} \qquad (A.5)$$

$$\beta = v/c \qquad (A.6)$$

If the velocity of a particle is $\mathbf{u}$, with components $u_x$, $u_y$, $u_z$ in frame $F$, its components in $F'$ are

[†] General reference, Rindler (1966).

$$u'_x = \frac{u_x - v}{1 - u_x v/c^2} \tag{A.7}$$

$$u'_y = \frac{u_y}{\gamma(1 - u_x v/c^2)} \tag{A.8}$$

$$u'_z = \frac{u_z}{\gamma(1 - u_x v/c^2)} \tag{A.9}$$

## A2  Four-vectors

The space-time coordinates of a point $(x, y, z, ct)$ are the components of a four-vector $x_\lambda$ ($\lambda = 1, 2, 3, 4$). In general a four-vector can be written as

$$(a_x, a_y, a_z, a_4) \quad \text{or} \quad (\mathbf{a}, a_4) \tag{A.10}$$

and transforms as

$$a'_x = \gamma(ax - \beta a_4) \tag{A.11}$$

$$a'_y = a_y \tag{A.12}$$

$$a'_z = a_z \tag{A.13}$$

$$a'_4 = \gamma(a_4 - \beta a_x) \tag{A.14}$$

under the Lorentz transformation, equations (A.1) to (A.4).

We also find it convenient at times to write the four-vector (A.10) as $a$.

The square of the scalar magnitude of a four-vector is

$$\begin{aligned}
a.a &= a_x^2 + a_y^2 + a_z^2 - a_4^2 \\
&= \mathbf{a}.\mathbf{a} - a_4^2
\end{aligned} \tag{A.15}$$

and is an invariant under Lorentz transformation.

For any two four-vectors, $a = (a_x, a_y, a_z, a_4)$, $b = (b_x, b_y, b_z, b_4)$, we define a scalar product

$$\begin{aligned}
a.b &= a_x b_x + a_y b_y + a_z b_z - a_4 b_4 \\
&= \mathbf{a}.\mathbf{b} - a_4 b_4
\end{aligned}$$

which is invariant under Lorentz transformations.

The momentum and energy of a particle form a four-vector $(\mathbf{p}, E/c)$ with square of the scalar magnitude $-M^2 c^2$.

$$p^2 - \frac{E^2}{c^2} = -M^2 c^2 \tag{A.16}$$

For any system of particles, the total momentum and total energy form a four-vector and

$$\left(\sum_i \mathbf{p}_i\right)^2 - \left(\frac{1}{c}\sum_i E_i\right)^2 \tag{A.17}$$

is invariant. Many problems can be dealt with simply by using the invariance of (A.17).

Note that $E$ is the total energy and includes the rest mass energy, i.e.

$$E = Mc^2 + T \tag{A.18}$$

where $T$ is the kinetic energy. Some useful relations for a particle with velocity $v$ are

$$p = \gamma Mc\beta \tag{A.19}$$

$$E = \gamma Mc^2$$
$$= c(p^2 + M^2 c^4)^{\frac{1}{2}} \tag{A.20}$$

$$\beta = cp/E \tag{A.21}$$

There is no generally accepted convention for writing four-vectors, and some confusion can occur when both four-vectors and ordinary three-vectors are used together. De Benedetti (1964) uses the notation

**k**, **a**, **A**, etc., are four-vectors

**k**, **a**, **A**, etc., are three-vectors

k, a, A, etc., are the magnitudes of four-vectors

$k, a, A$, etc., are the magnitudes of three-vectors

but unfortunately this notation is not in general use.

### A3 Transformation between laboratory frame and centre-of-mass frame

Consider a reaction

$$1 + 2 \rightarrow 3 + 4 \tag{A.22}$$

where particle 2 is initially at rest in the laboratory frame. In the centre-of-mass frame the total momentum is zero. We use unprimed quantities for laboratory coordinates and primed quantities for coordinates in the centre-of-mass frame. The total energy is

$$E_0 = M_2 c^2 + E_1 \tag{A.23}$$

From the invariance of (A.17),

$$p_1^2 - \frac{E_0^2}{c^2} = -\frac{E_0'^2}{c^2} \tag{A.24}$$

so that

$$E_0'^2 = M_1^2 c^4 + M_2^2 c^4 + 2M_2 E_1 c^2 \tag{A.25}$$

and

$$E_1 = (E_0'^2 - M_1^2 c^4 - M_2^2 c^4)/2M_2 c^2 \tag{A.26}$$

The centre-of-mass frame is moving with a velocity $v$ relative to the laboratory frame. Equation (A.21) can be applied to the two-particle system yielding

$$\beta = \frac{v}{c} = \frac{cp_1}{E_1 + M_2 c^2} \tag{A.27}$$

Note that

$$E_0 = \gamma E_0' \tag{A.28}$$

The energy released, or $Q$ value, of the reaction (A.22) is

$$Q = (M_1 + M_2 - M_3 - M_4)c^2 \tag{A.29}$$

If $Q$ is positive the reaction (A.22) can proceed for all values of $E_1$. If $Q$ is negative, the reaction has a threshold, i.e. there is a minimum value of $E_1$, say $E_1^{\text{th}}$, for which the reaction can occur. At threshold

$$E_0' = (M_3 + M_4)c^2 \tag{A.30}$$

and from (A.26)

$$E_1^{\text{th}} = \frac{[(M_3 + M_4)^2 - M_1^2 - M_2^2]c^2}{2M_2} \tag{A.31}$$

or

$$T_1^{\text{th}} = \frac{[(M_3 + M_4)^2 - (M_1 + M_2)^2]c^2}{2M_2} \tag{A.32}$$

We require the relation between the differential cross-sections in the laboratory and centre-of-mass frames. $(d\sigma/d\Omega)\, d\Omega$ is the probability that the particle considered is emitted from the reaction within a particular solid angle $d\Omega$. Since the probability of a particular event must be the same in all frames,

$$\frac{d\sigma}{d\Omega} d\Omega = \left(\frac{d\sigma}{d\Omega}\right)' d\Omega' \tag{A.33}$$

$$\frac{d\sigma}{d\Omega} = \left(\frac{d\sigma}{d\Omega}\right)' \frac{d\Omega'}{d\Omega} \tag{A.34}$$

$$d\Omega = d\phi \, d(\cos \theta) \tag{A.35}$$

where $\theta$ and $\phi$ are the angles of spherical polar coordinates. We take the $x$ direction along the direction of $\mathbf{p}_1$, and measure $\theta$ from the $x$ axis. $\phi$ is the azimuthal angle about the $x$ axis. Thus $\phi = \phi'$ and

$$\frac{d\sigma}{d\Omega} = \left(\frac{d\sigma}{d\Omega}\right)' \frac{d\cos \theta'}{d\cos \theta} \tag{A.36}$$

For the particle considered (either 3 or 4),

$$p'_x = \gamma(p_x - \beta E/c) \tag{A.37}$$

$$p'_y = p_y \tag{A.38}$$

$$p'_z = p_z \tag{A.39}$$

$$\frac{E'}{c} = \gamma\left(\frac{E}{c} - \beta p_x\right) \tag{A.40}$$

i.e.

$$p' \cos \theta' = \gamma(p \cos \theta - \beta E/c) \tag{A.41}$$

$$p' \sin \theta' = p \sin \theta \tag{A.42}$$

$$E' = \gamma(E - c\beta p \cos \theta) \tag{A.43}$$

In the centre-of-mass frame, the energy and magnitude of the momentum of the particle are independent of the direction of its motion, so that

$$\frac{dE'}{d\cos \theta} = 0 \quad \text{and} \quad \frac{dp'}{d\cos \theta} = 0 \tag{A.44}$$

Differentiating (A.41) with respect to $\cos \theta$ and using

$$\frac{dE}{dp} = \frac{c^2 p}{E} \tag{A.45}$$

(which follows from (A.20)), we obtain

$$p' \frac{d \cos \theta'}{d \cos \theta} = \gamma \left( p + \cos \theta \frac{dp}{d \cos \theta} - \beta \frac{cp}{E} \frac{dp}{d \cos \theta} \right) \quad \text{(A.46)}$$

and from (A.43)

$$0 = \gamma \left( \frac{c^2 p}{E} \frac{dp}{d \cos \theta} - c\beta p - c\beta \cos \theta \frac{dp}{d \cos \theta} \right) \quad \text{(A.47)}$$

Eliminating $dp/(d \cos \theta)$,

$$\frac{d \cos \theta'}{d \cos \theta} = \frac{p^2}{\gamma p' \left( p - \dfrac{\beta E}{c} \cos \theta \right)}$$

$$= \frac{p}{\gamma p' \left( 1 - \dfrac{\beta}{\beta_u} \cos \theta \right)} \quad \text{(A.48)}$$

where

$$\beta_u = u/c = cp/E$$

where $u$ is the velocity of the particle. Then

$$\frac{d\sigma}{d\Omega} = \frac{p}{\gamma p' \left( 1 - \dfrac{\beta}{\beta_u} \cos \theta \right)} \left( \frac{d\sigma}{d\Omega} \right)' \quad \text{(A.49)}$$

It is convenient to have the relation (A.48) also in terms of $\theta'$ and $E'$. Using the inverse of the Lorentz transformation which yielded equations (A.41) and (A.43),

$$p \cos \theta = \gamma \left( p' \cos \theta' + \frac{\beta E'}{c} \right) \quad \text{(A.50)}$$

$$E = \gamma (E' + c\beta p' \cos \theta') \quad \text{(A.51)}$$

Now

$$\frac{\cos \theta}{p} = \frac{p \cos \theta}{p^2} = \frac{1}{p^2} \gamma \left( p' \cos \theta + \frac{\beta E'}{c} \right) \quad \text{(A.52)}$$

From equations (A.52) and (A.51),

$$\frac{\beta E}{pc} \cos \theta = \frac{1}{p^2} \frac{\beta \gamma^2}{c} (E' + c\beta p' \cos \theta') \left( p' \cos \theta' + \frac{\beta E'}{c} \right)$$

so that

$$1 - \frac{\beta E}{pc} \cos \theta = \left[ p^2 - \frac{\beta \gamma^2}{c} (E' + c\beta p' \cos \theta') \left( p' \cos \theta' + \frac{\beta E'}{c} \right) \right] \bigg/ p^2$$

(A.53)

Substituting for $p^2$ in the numerator

$$p^2 = \frac{E^2}{c^2} - M^2 c^2 = \frac{E^2}{c^2} - \frac{E'^2}{c^2} + p'^2$$

$$= p'^2 - \frac{E'^2}{c^2} + \frac{\gamma^2}{c^2} (E' + c\beta p' \cos \theta')^2$$

(A.54)

we obtain

$$1 - \frac{\beta E}{pc} \cos \theta = \frac{p'^2}{p^2} \left( 1 + \frac{\beta E'}{p'c} \cos \theta' \right)$$

(A.55)

so that

$$\frac{d \cos \theta'}{d \cos \theta} = \frac{p^3}{\gamma p'^3 \left( 1 + \dfrac{\beta E'}{p'c} \cos \theta' \right)}$$

$$= \frac{p^3}{\gamma p'^3 \left( 1 + \dfrac{\beta}{\beta_u'} \cos \theta' \right)}$$

(A.56)

where

$$\beta_u' = u'/c.$$

(A.57)

Since the angle measured experimentally is $\theta$, and $p'$ and $E'$ are independent of angle, it is convenient to express $p$, $E$ and d $\cos \theta'$/d $\cos \theta$ in terms of $\theta$ and $p'$ or $\theta$ and $E'$.

From equation (A.43)

$$E = \frac{E'}{\gamma} + c\beta p \cos \theta$$

(A.58)

Then,

$$c^2 p^2 = E^2 - M^2 c^4$$

$$= \frac{E'^2}{\gamma^2} + \frac{2c\beta p E' \cos \theta}{\gamma} + c^2 \beta^2 p^2 \cos^2 \theta - M^2 c^4$$

i.e.

$$(1 - \beta^2 \cos^2 \theta) p^2 - 2 \frac{\beta E'}{c\gamma} p \cos \theta + M^2 c^2 - \frac{E'^2}{c^2 \gamma^2} = 0$$

Solving for $p$

$$p = \left[\frac{\beta E' \cos \theta}{c\gamma} \pm Y^{\frac{1}{2}}\right] \Big/ (1 - \beta^2 \cos^2 \theta) \qquad (A.59)$$

where

$$Y = \frac{E'^2}{c^2 \gamma^2} - M^2 c^2 (1 - \beta^2 \cos^2 \theta)$$
$$= p'^2 (1 - \beta^2) - M^2 c^2 \beta^2 \sin^2 \theta \qquad (A.60)$$

Equation (A.60) can be rearranged to yield

$$Y = \left(\frac{\beta E' \cos \theta}{c\gamma}\right)^2 + E'^2 (1 - \beta^2 \cos^2 \theta) \frac{\beta_u'^2 - \beta^2}{c^2} \qquad (A.61)$$

For

$$\beta_u' > \beta$$
$$Y > \left(\frac{\beta E' \cos \theta}{c\gamma}\right)^2$$

and since $p$ is the magnitude of the momentum and so must be positive, the positive sign must be taken in equation (A.59).
     For

$$\beta_u' < \beta$$

both signs can occur in equation (A.59), but for

$$\sin \theta > p'/\gamma M c \beta \qquad (A.62)$$

$Y$ is negative, and there is no real solution for $p$. In this case, where the centre-of-mass velocity is greater than the velocity with which the particle is emitted in the centre-of-mass frame, the particle cannot be emitted in backwards directions. For this case ($\beta_u' < \beta$) then

$$\sin \theta < p'/\gamma M c \beta \qquad (A.63)$$

and

$$\theta < \pi/2 \qquad (A.64)$$

and both signs occur in equation (A.59), as there are two possible values of $p$ at a particular $\theta$ corresponding to emission at two values of $\theta'$.
     By substituting equation (A.59) in

$$E = c(p^2 + M^2 c^2)^{\frac{1}{2}}$$

or, more easily, by writing equation (A.43) as

$$p = \frac{E - E'/\gamma}{c\beta \cos \theta} \tag{A.65}$$

and then substituting equation (A.43) for $p$ in

$$c^2 p^2 = E^2 - M^2 c^4$$

and solving the resulting quadratic equation for $E$, we obtain

$$E = \frac{E'/\gamma \pm c\beta \cos \theta \, Y^{\frac{1}{2}}}{(1 - \beta^2 \cos^2 \theta)} \tag{A.66}$$

From equations (A.66) and (A.59), we obtain

$$1 - \frac{\beta E}{cp} \cos \theta = \pm \frac{Y^{\frac{1}{2}} (1 - \beta^2 \cos^2 \theta)}{\dfrac{\beta E' \cos \theta}{c\gamma} \pm Y^{\frac{1}{2}}} \tag{A.67}$$

and equation (A.48) can then be written as

$$\frac{\mathrm{d} \cos \theta'}{\mathrm{d} \cos \theta} = \frac{\left( \dfrac{\beta E' \cos \theta}{c\gamma} \pm Y^{\frac{1}{2}} \right)^2}{\gamma p' Y^{\frac{1}{2}} (1 - \beta^2 \cos^2 \theta)^2} \tag{A.68}$$

Note that $Y$, which is given by equation (A.60), depends on $\theta$. The dependence of $\mathrm{d} \cos \theta'/\mathrm{d} \cos \theta$ on $\theta$ is then clearly shown by equations (A.68) and (A.60). The term $(1 - \beta^2 \cos^2 \theta)^2$ in the denominator in equation (A.68), shows that for a reaction at very large energy so that $\beta \simeq 1$, the final particles are produced predominantly in the direction of the incident particle. From equation (A.60) it is seen that $\mathrm{d} \cos \theta'/\mathrm{d} \cos \theta$ has a singularity at

$$\sin \theta = p'/Mc\beta\gamma$$

$$= \frac{\beta'_u}{(1 - \beta_u'^2)^{\frac{1}{2}}} \frac{(1 - \beta^2)^{\frac{1}{2}}}{\beta} \tag{A.69}$$

if $\beta'_u < \beta$, i.e. if the velocity with which the particle is emitted in the centre-of-mass frame is less than the velocity of the centre-of-mass. In this case, a peak occurs in the angular distribution of final particles in the laboratory frame at the angle $\theta$ given by equation (A.69). This feature of the relation (A.68) is used in the missing mass spectrometer for boson resonances discussed in section 45.

Note that equation (A.68) can also be written as

$$\frac{\mathrm{d} \cos \theta'}{\mathrm{d} \cos \theta} = \frac{p^2}{\gamma p' Y^{\frac{1}{2}}} \tag{A.70}$$

The results of this section apply also to the two-particle decay of a moving particle, say

$$5 \rightarrow 3+4$$

Then $v$ is the velocity of the decaying particle in the laboratory frame. The centre-of-mass frame is the rest frame of the decaying particle. In the case where the angular distribution of particles 3 and 4 is isotropic in the rest frame of 5, we see from equation (A.68) that, if particle 5 is extremely energetic so that $\beta \simeq 1$, then most of the decay particles are emitted towards the forwards direction in the laboratory frame. See Chapter 8, Exercises 2 and 3.

## A4 Time dilatation

An important result of the theory of special relativity is time dilatation, or the slowing down of moving clocks. A beam of decaying particles acts as a moving clock, and a beam of decaying particles decays slower than stationary particles. If $\tau_0$ is the mean lifetime (see Appendix C) of a particle in its rest frame, the observed mean lifetime in the laboratory frame in which the particle moves with velocity $v$ is

$$\tau = \frac{\tau_0}{(1-v^2/c^2)^{\frac{1}{2}}} \qquad \text{(A.71)}$$

(Rindler, 1966).

### *References*

DE BENEDETTI, S., *Nuclear Interactions*, 1964. Wiley, New York.

RINDLER, W., *Special Relativity*, 1966. Oliver and Boyd, London.

# Quantum mechanics

## B1 Introduction

Only a brief and rudimentary account of quantum mechanics is given here. Some suitable accounts of quantum mechanics are Feynman (1965), Ziock (1969), Saxon (1968), and at a more advanced level Schiff (1968), Landau and Lifshitz (1958).

To every free particle with momentum **p** and energy $E$, there is associated a de Broglie wave

$$\psi(\mathbf{r}, t) = \exp\left\{i(\mathbf{p}.\mathbf{r} - Et)/\hbar\right\} \tag{B.1}$$

The de Broglie wave has wavelength

$$\lambda = h/p \tag{B.2}$$

and frequency

$$\nu = E/h \tag{B.3}$$

The wave number is

$$k = 2\pi/\lambda = p/\hbar \tag{B.4}$$

In non-relativistic theory, $E$ is the kinetic energy of the particle

$$E = p^2/2M = \hbar^2 k^2/2M \tag{B.5}$$

and in relativistic theory, $E$ is the total energy of the particle

$$E = c(p^2 + M^2 c^2)^{\frac{1}{2}} \tag{B.6}$$

In the non-relativistic quantum mechanics of a system of $N$ particles, a physical state of the system is described by a wave function (or state function)

$$\psi(\mathbf{r}_1, \mathbf{r}_2, \ldots, \mathbf{r}_N, t)$$

which satisfies the Schrödinger equation

$$ i\hbar \frac{\partial}{\partial t} \psi(\mathbf{r}_1, \ldots, \mathbf{r}_N, t) = $$

$$ \left[ -\hbar^2 \sum_{i=1}^{N} \frac{1}{2M_i} \nabla_i^2 + V(\mathbf{r}_1, \ldots, \mathbf{r}_N) \right] \psi(\mathbf{r}_1, \ldots, \mathbf{r}_N, t) \quad \text{(B.7)} $$

where $V(\mathbf{r}_1, \ldots, \mathbf{r}_N)$ is the potential. For states of definite energy (also called stationary states)

$$ \psi(\mathbf{r}_1, \ldots, \mathbf{r}_N, t) = \psi_n(\mathbf{r}_1, \ldots, \mathbf{r}_N) \, e^{-iE_{nt}/\hbar} \quad \text{(B.8)} $$

where $\psi_n$ satisfies the time independent Schrödinger equation

$$ \left[ -\hbar^2 \sum_{i=1}^{N} \frac{1}{2M_i} \nabla_i^2 + V(\mathbf{r}_1, \ldots, \mathbf{r}_N) \right] \psi_n(\mathbf{r}_1, \ldots, \mathbf{r}_N) $$

$$ = E_n \psi_n(\mathbf{r}_1, \ldots, \mathbf{r}_N) \quad \text{(B.9)} $$

For a single particle

$$ i\hbar \frac{\partial}{\partial t} \psi(\mathbf{r}, t) = \left[ -\hbar^2 \frac{1}{2M} \nabla^2 + V(\mathbf{r}) \right] \psi(\mathbf{r}, t) \quad \text{(B.10)} $$

and for a state of definite energy

$$ \psi(\mathbf{r}, t) = \psi_n(\mathbf{r}) \, e^{-iE_{nt}/\hbar} $$
$$ \left[ -\hbar^2 \frac{1}{2M} \nabla^2 + V(\mathbf{r}) \right] \psi_n(\mathbf{r}) = E_n \psi_n(\mathbf{r}) \quad \text{(B.11)} $$

For a free particle, $V = 0$,

$$ i\hbar \frac{\partial}{\partial t} \psi(\mathbf{r}, t) = -\hbar^2 \frac{1}{2M} \nabla^2 \psi(\mathbf{r}, t) \quad \text{(B.12)} $$

of which the solution is given by equations (B.1) and (B.5).

## B2 States and operators

The quantum mechanics of an arbitrary physical system is obtained by abstracting some features of the non-relativistic quantum mechanics of a system of particles. Every state of a physical system is represented by a state function $\psi$. For a single spinless particle, $\psi$ can be taken as a function of the time and the coordinates of the particle, i.e. $\psi(\mathbf{r}, t)$, but more generally $\psi$ will depend on other quantities such as the spin. Quantities which can be measured or observed are called observables. To each observable $A$ of the system, there corresponds a linear operator $\hat{A}$. $\hat{A}$ is an operator if $\hat{A}\psi$ is a state function where $\psi$ is an arbitrary state function. $\hat{A}$ is linear if

$$\hat{A}(\lambda_1\psi_1 + \lambda_2\psi_2) = \lambda_1\hat{A}\psi_1 + \lambda_2\hat{A}\psi_2$$

for all state functions $\psi_1, \psi_2$ and all complex numbers $\lambda_1, \lambda_2$. Following Landau and Lifshitz (1958) and Feynman (1965), we use the notation $\hat{\ }$ to distinguish operators in this appendix. Some authors use a suffix, op, such as $A_{op}$, but more generally no distinguishing notation is used and it must be understood from the context whether $A$ stands for an observable or an operator. In the main text of this book, $\hat{\ }$ is used to distinguish operators only when confusion may arise otherwise.

Any two state functions $\psi_1$ and $\psi_2$ determine a complex number called the scalar product and written

$$\langle\psi_2|\psi_1\rangle$$

with the property

$$\langle\psi_1|\psi_2\rangle = \langle\psi_2|\psi_1\rangle^* \tag{B.13}$$

For state functions of a single particle without spin

$$\langle\psi_1|\psi_2\rangle = \int \psi_1^*\psi_2 \, d^3\mathbf{r} \tag{B.14}$$

When $\psi_1$ and $\psi_2$ are normalized so that

$$\langle\psi_1|\psi_1\rangle = 1$$
$$\langle\psi_2|\psi_2\rangle = 1$$

$|\langle\psi_2|\psi_1\rangle|^2$ is the probability that if the system is prepared in the state $\psi_1$, it will be observed to be in the state $\psi_2$.

$\hat{B}$ is called the Hermitean adjoint of $\hat{A}$ if

$$\langle\psi_1|\hat{B}\psi_2\rangle = \langle\psi_2|\hat{A}\psi_1\rangle^*$$

for all $\psi_1, \psi_2$, and is written as

$$\hat{B} = \hat{A}^\dagger$$

An operator $\hat{A}$ is called a hermitean operator if

$$A^\dagger = \hat{A}$$

Real observables, such as momentum and energy, are represented by hermitean operators. For any hermitean operator $\hat{A}$, a set of state functions $\psi_n$, called the eigenfunctions of $\hat{A}$, can be found such that

$$\hat{A}\psi_n = a_n\psi_n \tag{B.15}$$

where the $a_n$ are real numbers, called the eigenvalues of $\hat{A}$. The states corresponding to the $\psi_n$ are called the eigenstates of $\hat{A}$. The result of a measurement of an observable $A$ is one of the eigenvalues $a_n$. The average value of a large number of measurements of an observable $A$ carried out on a system described by a state function $\psi$ is

$$\langle A \rangle = \frac{\langle \psi | \hat{A} \psi \rangle}{\langle \psi | \psi \rangle} \qquad (B.16)$$

In general, operators do not commute

$$\hat{A}\hat{B} - \hat{B}\hat{A} \neq 0$$

e.g. for a single particle described by state functions $\psi(\mathbf{r}, t)$, the components of momentum are represented by the operators

$$\hat{p}_x = \frac{\hbar}{i} \frac{\partial}{\partial x}$$

$$\hat{p}_y = \frac{\hbar}{i} \frac{\partial}{\partial y} \qquad (B.17)$$

$$\hat{p}_z = \frac{\hbar}{i} \frac{\partial}{\partial z}$$

and position coordinates $x, y, z$ by operators $\hat{x} = x$, $\hat{y} = y$, $\hat{z} = z$, i.e. the operator $\hat{x}$ is just multiplication by $x$. Then

$$\hat{p}_x \hat{x} - \hat{x}\hat{p}_x = \hbar/i$$
$$\hat{p}_y \hat{y} - \hat{y}\hat{p}_y = \hbar/i \qquad (B.18)$$
$$\hat{p}_z \hat{z} - \hat{z}\hat{p}_z = \hbar/i$$

The notation

$$[\hat{A}, \hat{B}] = \hat{A}\hat{B} - \hat{B}\hat{A}$$

is generally used for commutators.

## B3  Angular momentum

The orbital angular momentum operator for a single particle is

$$\hat{\mathbf{M}} = \hbar\hat{\mathbf{L}} = \hat{\mathbf{r}} \times \hat{\mathbf{p}} \qquad (B.19)$$

and has the commutation relations

$$[\hat{L}_x, \hat{L}_y] = i\hat{L}_z, [\hat{L}_y, \hat{L}_z] = i\hat{L}_x, [\hat{L}_z, \hat{L}_x] = i\hat{L}_y \qquad (B.20)$$

For a spherically symmetric potential, $V(\mathbf{r}) = V(r)$, and the solution of equation (B.11) can be written as

$$\psi_n(\mathbf{r}) = R_{nl}(r) Y_{lm}(\theta, \phi) \tag{B.21}$$

where $r$, $\theta$, $\phi$ are spherical polar coordinates

$$x = r \sin \theta \cos \phi$$
$$y = r \sin \theta \sin \phi$$
$$z = r \cos \theta$$

and $Y_{lm}(\theta, \phi)$ is the spherical harmonic

$$Y_{lm}(\theta, \phi) = (-1)^m \left[ \frac{(2l+1)(l-m)!}{4\pi(l+m)!} \right]^{\frac{1}{2}} e^{im\phi} P_l^{|m|}(\cos \theta) \tag{B.22}$$

where $P_l^{|m|}(z)$ are the associated Legendre polynomials. $l$ is the orbital angular momentum quantum number of the particle and $m$ is the quantum number for the $z$-component of orbital angular momentum. The $Y_{lm}(\theta, \phi)$ are eigenfunctions of the $z$ component of orbital angular momentum and of the magnitude of the orbital angular momentum,

$$\hat{M}_z Y_{lm}(\theta, \phi) = \hbar m Y_{lm}(\theta, \phi) \tag{B.23}$$

$$\hat{M}^2 Y_{lm}(\theta, \phi) = \hbar^2 \hat{L}^2 Y_{lm}(\theta, \phi) = \hbar^2 l(l+1) Y_{lm}(\theta, \phi) \tag{B.24}$$

$l$ is a non-negative integer, and $m$ is an integer

$$-l \leqslant m \leqslant +l$$

When spin is included, the angular momentum quantum numbers may be half integral. All angular momentum operators $\hbar \hat{\mathbf{J}}$ obey the commutation relations

$$[\hat{J}_x, \hat{J}_y] = i\hat{J}_z, [\hat{J}_y, \hat{J}_z] = i\hat{J}_x, [\hat{J}_z, \hat{J}_x] = i\hat{J}_y \tag{B.25}$$

and the eigenfunctions of $\hat{J}^2$ can also be eigenfunctions of one component of $\hat{\mathbf{J}}$, usually chosen as $\hat{J}_z$,

$$\hat{J}^2 \psi_{JM} = J(J+1) \psi_{JM} \tag{B.26}$$

$$\hat{J}_z \psi_{JM} = M \psi_{JM} \tag{B.27}$$

where $J$ is a non-negative integer or half integer, and $M$ takes the $(2J+1)$ values

$$M = -J, -J+1, \ldots, J-1, J$$

## B4 Addition of angular momenta

We consider adding two angular momenta $\hat{\mathbf{J}}_a, \hat{\mathbf{J}}_b$ to give a total angular momentum $\hat{\mathbf{J}}$,

$$\hat{\mathbf{J}} = \hat{\mathbf{J}}_a + \hat{\mathbf{J}}_b \tag{B.28}$$

For example, $\hat{\mathbf{J}}_a, \hat{\mathbf{J}}_b$ may be the orbital angular momenta of two different particles, or the orbital angular momentum and spin of a single particle. $\hat{\mathbf{J}}_a, \hat{\mathbf{J}}_b$ refer to independent systems so that

$$[\hat{\mathbf{J}}_a, \hat{\mathbf{J}}_b] = 0 \tag{B.29}$$

Let $\psi^a_{J_a M_a}$ and $\psi^b_{J_b M_b}$ be eigenstates of $\hat{J}_a^2, \hat{J}_{az}$ and $\hat{J}_b^2, \hat{J}_{bz}$ respectively.

$$\begin{aligned}
\hat{J}_a^2 \psi^a_{J_a M_a} &= J_a(J_a+1)\psi^a_{J_a M_a} \\
\hat{J}_{az} \psi^a_{J_a M_a} &= M_a \psi^a_{J_a M_a} \\
\hat{J}_b^2 \psi^b_{J_b M_b} &= J_b(J_b+1)\psi^b_{J_b M_b} \\
\hat{J}_{bz} \psi^b_{J_b M_b} &= M_b \psi^b_{J_b M_b}
\end{aligned} \tag{B.30}$$

The products $\psi^a_{J_a M_a} \psi^b_{J_b M_b}$ are eigenfunctions of $\hat{J}_a^2, \hat{J}_b^2, \hat{J}_{az}, \hat{J}_{bz}$. We require the eigenfunctions of $\hat{J}_a^2, \hat{J}_b^2, \hat{J}^2, \hat{J}_z$,

$$\begin{aligned}
\hat{J}^2 \psi_{JM} &= J(J+1)\psi_{JM} \\
\hat{J}_z \psi_{JM} &= M\psi_{JM} \\
\hat{J}_a^2 \psi_{JM} &= J_a(J_a+1)\psi_{JM} \\
J_b^2 \psi_{JM} &= J_b(J_b+1)\psi_{JM}
\end{aligned} \tag{B.31}$$

The $\psi_{JM}$ are given by

$$\psi_{JM} = \sum_{M_a, M_b} \langle J_a M_a J_b M_b | J_a J_b J M \rangle \psi^a_{J_a M_a} \psi^b_{J_b M_b} \tag{B.32}$$

where the $\langle J_a M_a J_b M_b | J_a J_b J M \rangle$ are called the Clebsch–Gordan coefficients and, as the notation implies, are the scalar products

$$\langle J_a M_a J_b M_b | J_a J_b J M \rangle = \langle \psi^a_{J_a M_a} \psi^b_{J_b M_b} | \psi_{JM} \rangle \tag{B.33}$$

Note that

$$\langle J_a M_a J_b M_b | J_a J_b J M \rangle = 0 \tag{B.34}$$

for $M \neq M_a + M_b$.

A convenient notation is obtained by writing a state function $\psi_a$ as $|a\rangle$, where $a$ is the set of quantum numbers or labels of the state, e.g. we replace $\psi_{JM}$ by $|JM\rangle$. Where a state function, say $\psi_b$, appears as a prefactor in a scalar product, such as $\langle \psi_b | \psi_a \rangle$ it is replaced by $\langle b|$. Then

$$\langle \psi_b | \psi_a \rangle \equiv \langle b | a \rangle$$

$\langle b|$ is called a bra, and $|a\rangle$ is called a ket; the two together forming a bracket $\langle b|a \rangle$. The use of bras and kets has more significance than just convenience of notation, but we shall make no use of their deeper significance. For further treatment of bras and kets see Feynman (1965) and Dirac (1958).

Using kets, equations (B.26) and (B.27) can be written

$$\hat{J}^2 |J, M\rangle = J(J+1)|J, M\rangle$$
$$\hat{J}_z |J, M\rangle = M|J, M\rangle \tag{B.35}$$

Equation (B.32) can be written

$$|J_a J_b JM\rangle = \sum_{M_a, M_b} \langle J_a M_a J_b M_b | J_a J_b JM \rangle |J_a M_a J_b M_b\rangle \tag{B.36}$$

The inverse of equation (B.36) is

$$|J_a M_a J_b M_b\rangle = \sum_{J, M} \langle J_a J_b JM | J_a M_a J_b M_b \rangle |J_a J_b JM\rangle \tag{B.37}$$

and

$$\langle J_a J_b JM | J_a M_a J_b M_b \rangle = \langle J_a M_a J_b M_b | J_a J_b JM \rangle^* \tag{B.38}$$

The relative phases of angular momentum state functions are usually chosen so that the Clebsch–Gordan coefficients are real.

$$\langle J_a M_a J_b M_b | J_a J_b JM \rangle \neq 0$$

only for

$$J = J_a + J_b, J_a + J_b - 1, \ldots, |J_a - J_b| \tag{B.39}$$

so that the coupling of angular momenta is as described by the vector model (Eisberg, 1961). Using equations (B.39) and (B.34), equation (B.37) can be written

$$|J_a M_a J_b M_b\rangle = \sum_{J = J_a - J_b}^{J_a + J_b} \langle J_a J_b J(M_a + M_b) | J_a M_a J_b M_b \rangle$$
$$\times |J_a J_b J(M_a + M_b)\rangle \tag{B.40}$$

Clebsch–Gordan coefficients for $J_b = \frac{1}{2}$ and arbitrary $J_a$ are given in Table B.1.

TABLE B.1   $\langle J_a(M-M_b)\tfrac{1}{2}M_b|J_a\tfrac{1}{2}JM\rangle$

| $J$ | $M_b = +\tfrac{1}{2}$ | $M_b = -\tfrac{1}{2}$ |
|---|---|---|
| $J_a+\tfrac{1}{2}$ | $\left[\dfrac{J_a+M+\tfrac{1}{2}}{2J_a+1}\right]^{\frac{1}{2}}$ | $\left[\dfrac{J_a-M+\tfrac{1}{2}}{2J_a+1}\right]^{\frac{1}{2}}$ |
| $J_a-\tfrac{1}{2}$ | $-\left[\dfrac{J_a-M+\tfrac{1}{2}}{2J_a+1}\right]^{\frac{1}{2}}$ | $\left[\dfrac{J_a+M+\tfrac{1}{2}}{2J_a+1}\right]^{\frac{1}{2}}$ |

### References

DIRAC, P. A. M., *The Principles of Quantum Mechanics*, 4th edition, 1958. Oxford University Press.

EISBERG, R. M., *Fundamentals of Modern Physics*, 1961. Wiley, New York.

FEYNMAN, R. P., R. B. LEIGHTON and M. SANDS, *Quantum Mechanics*, Vol. III of *The Feynman Lectures on Physics*, 1965. Addison-Wesley, Reading, Mass.

LANDAU, L. D. and E. M. LIFSHITZ, *Quantum Mechanics*, 1958. Pergamon, London.

SAXON, D. S., *Quantum Mechanics*, 1968. Holden-Day, San Francisco.

SCHIFF, L. I., *Quantum Mechanics*, 3rd edition, 1968. McGraw-Hill, New York.

ZIOCK, K., *Basic Quantum Mechanics*, 1969. Wiley, New York.

## C1 Lifetime

If the probability per unit time of a particle decaying is $\lambda$, and if there are $N$ particles of which $dN$ decay in time $dt$, then

$$dN = -N\lambda\, dt \tag{C.1}$$

Hence

$$N = N_0 \exp(-\lambda t) \tag{C.2}$$

where $N_0$ is the number of particles at the beginning $t=0$. The *lifetime* $\tau$ of the decaying particles is defined as the reciprocal of the decay probability per unit time,

$$\tau = 1/\lambda \tag{C.3}$$

Then

$$N = N_0 \exp(-t/\tau) \tag{C.4}$$

The lifetime is therefore the time in which the number of particles is reduced to $1/e$ of the value it had at the beginning of the time interval.

The lifetime defined by equation (C.3) is the mean lifetime, for the mean lifetime is given by

$$\frac{\int t \exp(-t/\tau)\, dt}{\int \exp(-t/\tau)\, dt} = \tau \tag{C.5}$$

Although the mean lifetime is the most commonly used measure of decay probability in particle physics, some authors (e.g. Livingston, 1968) also use the half-life $T$, which is defined as the time in which the number of particles is reduced to the fraction $\frac{1}{2}$ of the value it had at the beginning of the time interval, and which is related to the lifetime by

$$T = \tau \ln 2 = 0.693\,\tau$$

## C2 Cross-section

The probability of a reaction is usually given by its cross-section. Consider a reaction

$$A + B \rightarrow \text{anything} \tag{C.6}$$

where a beam of particles $A$ of well-defined momentum is incident on a target of particles $B$, where the probability of an $A$ particle reacting with a $B$ particle is small enough so that attenuation of the beam of $A$ particles through the target of $B$ particles is negligible. The cross-section (or total cross-section) is defined as the number of events, of type given by equation (C.6), occurring per unit time per unit flux of incident particles per target particle. The flux of the beam of incident particles is the number of particles crossing unit area perpendicular to the direction of the beam per unit time. For the situation depicted in Fig. C.1, where $N_A$ particles cross the area $S$ and $N_E$ events occur in time $t$ for a target containing $N_B$ particles, the cross-section is

$$\sigma = \frac{N_E/t}{N_B N_A t^{-1} S^{-1}} = \frac{N_E S}{N_B N_A} \tag{C.7}$$

and the cross-section has the dimensions of area.

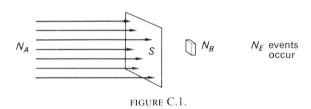

FIGURE C.1.

When the direction of motion of some particle, say $C$, in the final state is measured, a differential cross-section $d\sigma/d\Omega$ can be determined. The number of events per unit time per unit incident flux per target particle such that particle $C$ has direction of motion within a specified solid angle of $d\Omega$ is

$$\frac{d\sigma}{d\Omega} \, d\Omega$$

### Reference

LIVINGSTON, M. S., *Particle Physics: The High-Energy Frontier*, 1968. McGraw-Hill, New York.

# Principle of detailed balance

Consider the reaction

$$A + a \rightleftarrows B + b \tag{D.1}$$

and denote the channels $A + a$ and $B + b$ by $\alpha$ and $\beta$ respectively.

Imagine the particles $A$, $a$, $B$, $b$ to occur in arbitrary numbers in a large box of volume $\Omega$. At equilibrium, the number of transitions $\alpha \to \beta$ per unit time is equal to the number of transitions $\beta \to \alpha$ per unit time. The principle of detailed balance asserts that this balance of transition rates holds in detail for the energy range $(E, E + dE)$.

Suppose there are $N_\alpha$ states of channel $\alpha$ and $N_\beta$ states of channel $\beta$ in the energy range $(E, E + dE)$. According to the principles of statistical mechanics, at equilibrium all states in the range $(E, E + dE)$ have equal probability of being occupied. Then the number of states occupied in channels $\alpha$ and $\beta$ will be proportional to $N_\alpha$ and $N_\beta$ respectively.

Denoting the average transition probability per unit time from a state of $\alpha$ in the range $(E, E + dE)$ to the whole range of states of $\beta$ in $(E, E + dE)$ by $w(\alpha \to \beta)$, and the corresponding transition probability per unit time for transitions $\beta \to \alpha$ by $w(\beta \to \alpha)$, the balance of transition rates for the range $(E, E + dE)$ is

$$N_\alpha w(\alpha \to \beta) = N_\beta w(\beta \to \alpha) \tag{D.2}$$

It is convenient to work in the centre-of-mass frame, defined by

$$\mathbf{p}_A + \mathbf{p}_a = 0 = \mathbf{p}_B + \mathbf{p}_b \tag{D.3}$$

From quantum mechanics, for spinless particles, there is one state per volume $h^3$ of phase space. Thus in a volume $\Omega$, there are

$$4\pi h^{-3} p_a^2 \, dp_a \Omega$$

states of $a$ in the range $(p_a, p_a + dp_a)$. Since we are working in the centre-of-mass frame, the state of $a$ uniquely determines the state of $A$, so that

$$N_\alpha = 4\pi h^{-3} p_a^2 \, dp_a \Omega \tag{D.4}$$

Writing

$$p_a = p_A = p_\alpha \tag{D.5}$$

then

$$E = E_a + E_A \tag{D.6}$$

where

$$E_a = c(p_\alpha^2 + M_a^2 c^2)^{\frac{1}{2}}$$
$$E_A = c(p_\alpha^2 + M_A^2 c^2)^{\frac{1}{2}} \tag{D.7}$$

Then

$$\begin{aligned} dE &= \frac{c^2 p_\alpha}{E_a} \, dp_\alpha + \frac{c^2 p_\alpha}{E_A} \, dp_\alpha \\ &= (v_a + v_A) \, dp_\alpha \\ &= V_\alpha \, dp_\alpha \end{aligned} \tag{D.8}$$

where $\mathbf{V}_\alpha$ is the rate of change of $\mathbf{r}_a - \mathbf{r}_A$ computed in the centre-of-mass system.

So that

$$N_\alpha = 4\pi h^{-3} \Omega p_\alpha^2 V_\alpha^{-1} \, dE \tag{D.9}$$

and similarly

$$N_\beta = 4\pi h^{-3} \Omega p_\beta^2 V_\beta^{-1} \, dE \tag{D.10}$$

Substituting (D.9) and (D.10) into equation (D.2), we find

$$p_\alpha^2 V_\alpha^{-1} w(\alpha \to \beta) = p_\beta^2 V_\beta^{-1} w(\beta \to \alpha) \tag{D.11}$$

The situations we have just considered correspond to an incident flux of $V\Omega^{-1}$ per unit area per unit time, with $V = V_\alpha$ or $V = V_\beta$, depending on whether we are considering the forward or reverse reaction. The total cross-section $\sigma$ is defined as the number of transitions per unit time per unit incident flux, so that

$$\sigma = \Omega w / V \tag{D.12}$$

and equation (D.11) can be written

$$p_\alpha^2 \sigma(\alpha \to \beta) = p_\beta^2 \sigma(\beta \to \alpha) \tag{D.13}$$

If a particle has spin $J$, some component of the spin, say $J_z$, must be specified to completely specify the state of the particle, and each spatial state corresponds to $2J+1$ states. Thus the corresponding density of states must be multiplied by $2J+1$. If the particles $A$, $a$, $B$, $b$ have spins $J_A$, $J_a$, $J_B$, $J_b$ respectively, equation (D.13) must be modified to

$$(2J_A+1)(2J_a+1)p_\alpha^2 \sigma(\alpha \to \beta) = (2J_B+1)(2J_b+1)p_\beta^2 \sigma(\beta \to \alpha)$$
(D.14)

We have assumed that the incident beam of particles and the target are unpolarized, and that the polarization of the final state is not measured. If the cross-sections are measured for particular spin states, as in a polarization experiment, equation (D.13) must be used.

In calculating the density of states, it was assumed that particles $a$ and $A$ are distinguishable, and similarly for particles $b$ and $B$. If, for instance, particles $a$ and $A$ are the same, so that the reaction is

$$a+a \rightleftarrows B+b$$
(D.15)

the initial quantum mechanical state must be symmetric or anti-symmetric with respect to interchange of the two $a$ particles, according to whether they are bosons or fermions, and only half the states we have included can physically occur. In this case we have

$$\tfrac{1}{2}(2J_a+1)^2 p_\alpha^2 \sigma(\alpha \to \beta) = (2J_b+1)(2J_B+1)p_\beta^2 \sigma(\beta \to \alpha)$$
(D.16)

# Resonance of classical oscillator

The equation of motion of a damped harmonic oscillator is

$$m\frac{d^2x}{dt^2}+\beta\frac{dx}{dt}+kx = 0 \tag{E.1}$$

The solution to this equation is of the form

$$x = a\,e^{\lambda_1 t}+b\,e^{\lambda_2 t} \tag{E.2}$$

where $\lambda_1$ and $\lambda_2$ are the two roots of

$$m\lambda^2+\beta\lambda+k = 0 \tag{E.3}$$

i.e.

$$\begin{aligned}\lambda_1 &= \frac{-\beta}{2m}+\left[\frac{\beta^2}{4m^2}-\frac{k}{m}\right]^{\frac{1}{2}} \\[2mm] \lambda_2 &= \frac{-\beta}{2m}-\left[\frac{\beta^2}{4m^2}-\frac{k}{m}\right]^{\frac{1}{2}}\end{aligned} \tag{E.4}$$

There are two types of solution:

(a) If $\beta^2 \geqslant 4\,km$, then the $\lambda$ are real and there are no oscillations. We are not interested in this case.
(b) If $\beta^2 < 4\,km$, the solution is oscillatory with an exponentially decreasing amplitude.

$$x = e^{-(\beta/2m)t}\left[a\,e^{i\omega t}+b\,e^{-i\omega t}\right] \tag{E.5}$$

where

$$\omega = \left(\frac{k}{m}-\frac{\beta^2}{4m^2}\right)^{\frac{1}{2}} \tag{E.6}$$

Alternatively, the solution can be written as

$$x = c\,e^{-(\beta/2m)t}\cos(\omega t-\gamma) \tag{E.7}$$

and an example is sketched in Fig. E.1.

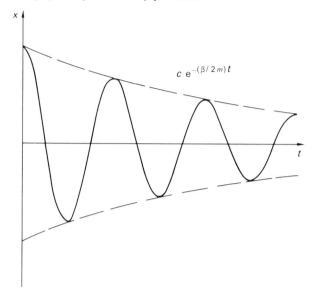

FIGURE E.1    Variation of amplitude with time for a damped oscillator.

The damped frequency

$$v = \frac{1}{2\pi} \left[ \frac{k}{m} - \frac{\beta^2}{4m^2} \right]^{\frac{1}{2}}$$    (E.8)

is somewhat less than the undamped frequency

$$v_0 = \frac{1}{2\pi} \left[ \frac{k}{m} \right]^{\frac{1}{2}}$$    (E.9)

The square of the amplitude of the damped motion decreases as $\exp[-(\beta/m)t]$. We thus identify $m/\beta$ with the mean lifetime of the decaying oscillation,

$$\tau = m/\beta$$    (E.10)

Consider now the effect of a driving force $F_0 e^{i\omega t}$. The equation of motion for the forced vibration is

$$m \frac{d^2 x}{dt^2} + \beta \frac{dx}{dt} + kx = F_0 e^{i\omega t}$$    (E.11)

The solution for steady oscillations is

$$x = a e^{i(\omega t - \phi)}$$    (E.12)

with $a$ and $\phi$ real. Substituting (E.12) in (E.11), we get

$$a\left(\omega_0^2 - \omega^2 + i\frac{\beta\omega}{m}\right) = \frac{F_0}{m}e^{i\phi} \tag{E.13}$$

where

$$\omega_0 = \left(\frac{k}{m}\right)^{\frac{1}{2}} \tag{E.14}$$

Taking the square of the modulus of equation (E.13),

$$a^2\left[(\omega^2 - \omega_0^2)^2 + \frac{\beta^2\omega^2}{m^2}\right] = \left(\frac{F_0}{m}\right)^2$$

$$\therefore \frac{a^2}{F_0^2} = \frac{1}{m^2}\frac{1}{[\{(\omega-\omega_0)(\omega+\omega_0)\}^2 + \beta^2\omega^2/m^2]} \tag{E.15}$$

For $\omega$ near $\omega_0$,

$$\frac{a^2}{F_0^2} \simeq \frac{1}{4\omega_0^2 m^2}\frac{1}{(\omega-\omega_0)^2 + \beta^2/4m^2} \tag{E.16}$$

From equation (E.15) (or equation (E.16)) we see that the response of the oscillator is greatest for $\omega = \omega_0$, when the driving frequency is the same as the natural frequency of the oscillator.

To make an analogy with quantum theory, we write

$$E = \hbar\omega, \qquad E_0 = \hbar\omega_0$$

and then equation (E.16) can be written

$$\frac{a^2}{F_0^2} = \frac{\hbar^2}{4\omega_0^2 m^2}\frac{1}{(E-E_0)^2 + \hbar^2\beta^2/4m^2} \tag{E.17}$$

The second factor has the form

$$\frac{1}{(E-E_0)^2 + \Gamma^2/4} \tag{E.18}$$

with

$$\Gamma = \hbar^2\beta^2/m^2 \tag{E.19}$$

The expression (E.18) has a maximum of $4/\Gamma^2$ at $E = E_0$ and a total width at half maximum of $\Gamma$. From equations (E.19) and (E.10), we see that

$$\Gamma\tau = \hbar \tag{E.20}$$

There is a definite relationship for a classical oscillator between the width of the resonance under forced vibrations and the lifetime of the free motion. For the corresponding quantum system, this relation is the Heisenberg uncertainty relation.

# Experimental methods of high-energy physics

## F1 Introduction

An adequate account of experimental high-energy physics would require more space than the whole of this book, and it is only possible in this appendix to give a very brief account which can act as a guide to the references listed. For a general account of experimental high-energy physics, see Segrè (1964) and at a very elementary level, Gouiran (1967).

## F2 Particle accelerators

Experiments in particle physics involve a source of particles, a means of detection, and frequently extremely complicated equipment for handling the beam. In early work, the source was cosmic-ray particles, and this remains the only source of extremely energetic particles. Accelerators provide particles of energy up to 200 GeV. The types of accelerator used for particle physics are listed in Table F.1. The various types are differentiated mainly by the method of applying the electric field which accelerates the particles. All high-energy accelerators use radiofrequency electric fields that are applied many times to the particles. In a linear accelerator the application of the radiofrequency field is made at many places along a straight path. In circular accelerators, such as synchrocyclotrons and synchrotrons, the particles are confined to circular or spiralling paths by magnetic guide fields so that they pass one or several radiofrequency sources a large number of times.

Synchrotrons usually have straight sections to provide space for the radiofrequency accelerating system and auxiliary equipment, and for extraction of the main beam or of secondary beams. By allowing the main beam to strike either an internal or external target, it is possible to generate secondary beams of such particles as pions, kaons and antiprotons. Beam sharing techniques allow the performance of several experiments simultaneously.

TABLE F.1    Types of Accelerators used for High-Energy Physics

| Accelerator type | Particle orbit | Magnetic field | Method of acceleration | Maximum energy (1972) |
|---|---|---|---|---|
| Proton synchro-cyclotron | Circle of in-creasing radius | Constant in time | RF of decreas-ing frequency (about 30–20 MHz) | 1 GeV |
| Proton synchro-tron | Circle of con-stant radius | Increasing in time | RF of increas-ing frequency with large ranges | 300 GeV |
| Electron synchrotron | Circle of in-creasing radius | Increasing in time | RF of constant frequency (40–700 MHz) | 10 GeV |
| Electron linear accelerator | Straight line | None | RF of constant frequency (1300–3000 MHz) | 21 GeV |

Introductory accounts of accelerators are given by Wilson (1958) and Gourian (1967). For further information, see Rosenblatt (1968). More specialized accounts are given by Livingston and Blewett (1962), Blewett (1967, 1969), Courant (1968).

## F3  Intersecting storage rings

For a particle of mass $M_1$ incident on a stationary target particle of mass $M_2$, the energy available for the production of additional particles is

$$W = E_{c.m.} - (M_1 + M_2)c^2 = (E_{lab}^2 - c^2 p_1^2)^{\frac{1}{2}} - (M_1 + M_2)c^2$$

where $E_{c.m.}$ and $E_{lab}$ are the total energy in the centre-of-mass frame and laboratory frame respectively, and $p_1$ is the momentum in the laboratory frame of the accelerated particle. Then for non-relativistic motion $p_1 \ll M_1 c$, $p_1 \ll M_2 c$,

$$W \simeq \frac{M_2}{M_1 + M_2} T_1$$

where $T_1$ is the kinetic energy in the laboratory frame of the incident particle. For extremely relativistic motion, $p_1 \gg M_1 c$, $p_2 \gg M_2 c$,

$$W \simeq c(2M_2 T_1)^{\frac{1}{2}}$$

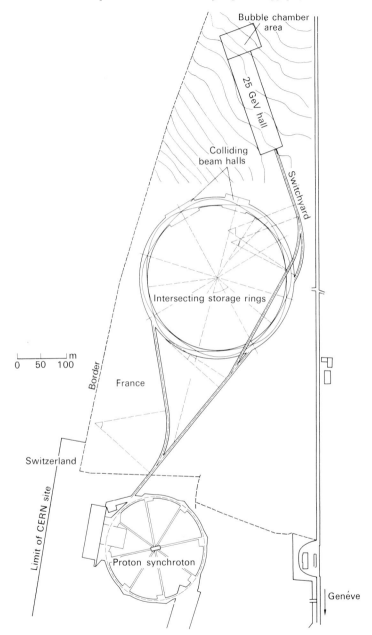

FIGURE F.1 General plan of the CERN Intersecting Storage Rings facility.
(From Blewett, 1969.)

i.e.

$$\frac{W}{T_1} \simeq c \left(\frac{2M_2}{T_1}\right)^{\frac{1}{2}}$$

and a decreasing proportion of the incident kinetic energy is available for particle production. For instance, for 30 GeV protons incident on protons at rest, only about 6 GeV is available for the production of particles. However, if two 30 GeV protons were to collide head on, 60 GeV would be available for particle production. Thus, much larger effective energies can be obtained by using colliding beams of particles, and for this reason many storage rings have been constructed for colliding beam experiments, for electron–electron, electron–positron, proton–proton and even proton–antiproton colliding beams. In some colliding beam arrangements, particles are accelerated and stored in the same ring, and in other arrangements, such as the CERN Intersecting Storage Rings (ISR) facility, the

FIGURE F.2 Intersection point at the CERN Intersecting Storage Rings. Here the tunnel is widened on the inside (left) by 3 m, and the floor is lowered by 2·4 m to accommodate experimental equipment. At centre right is the end of one of the beam transfer tunnels, which brings protons from the synchrotron to the ISR. The injection point for one of the rings is at the next crossing point, 100 m off the bottom of the photograph. (From *Physics Today*, April 1971.)

storage rings are separate from the accelerator. The CERN ISR are shown in Figs. F.1 and F.2. The first observation of colliding beam events in the CERN ISR occurred early in 1971 using 15 GeV protons (ISR staff, 1971). When operated at the full energy of 28 GeV, the collisions correspond to a beam of 1700 GeV hitting a stationary target.

An introductory account of storage rings is given by O'Neill (1966). For more detailed accounts, see Courant (1968), Blewett (1969).

## F4  Particle detectors

All methods of particle detection depend on observing effects of the charge of particles, so that only charged particles can be detected directly. The presence of neutral particles can only be inferred from their interactions with charged particles or from their charged decay products. Apart from Cerenkov counters, all other methods of particle detection depend on the ionization produced by charged particles passing through matter. Particle detectors can be divided into two classes: counters, in which a light pulse or electrical pulse indicates that a particle has gone through the detector, and track detectors in which a picture is formed of the paths followed by charged particles.

The most important counters for high-energy physics are scintillation counters and Cerenkov counters. In a scintillation counter, a flash of light results from atomic transitions in the material of the counter, after ionization has been caused by a fast charged particle. The light is converted by a photomultiplier to a sufficiently large electrical pulse to act as input to electronic apparatus used to analyse experiments.

Cerenkov counters make use of the light produced as an electromagnetic shock wave when a charged particle traverses a transparent medium with a velocity greater than the velocity of light in the material. The velocity of light in a medium of refractive index $n$ is $c/n$. The directions of emission of Cerenkov radiation from a particle with velocity $\beta c$ lie in a cone of semiangle $\theta$ about the direction of motion of the particle, where

$$\cos \theta = 1/\beta n$$

Thus it is possible to distinguish between particles of different velocity by choosing the refractive index of the medium and the range of angles for which the light is observed. As in the scintillation counter, the light is collected by photomultipliers.

For an account of the use of counters in particle physics see Gibson (1970). Detailed accounts are given by Birks (1964) of scintillation counters, and by Hutchinson (1960) of Cerenkov counters.

Of the track detectors, the Wilson cloud chamber (Henderson, 1970) and the photographic emulsion (Powell *et al.*, 1959) were the main methods used in particle physics for many years, but are now used only for special purposes. In the cloud chamber, the ions marking the path of a charged particle act as centres for condensation of water drops from a supersaturated vapour. In photographic emulsion, ionization causes grains of silver to become developable to metallic silver.

The two most important track detectors at present are the bubble chamber and the spark chamber.

### F5  Bubble chambers

The bubble chamber was invented by Glaser in 1952 (see Glaser, 1955). In the bubble chamber, a trail of bubbles is formed in a superheated liquid by local heating along the path of a charged particle. The superheated condition of the liquid is achieved by a rapid pressure drop starting from an equilibrium pressure high enough that boiling cannot occur. The tracks of bubbles are photographed

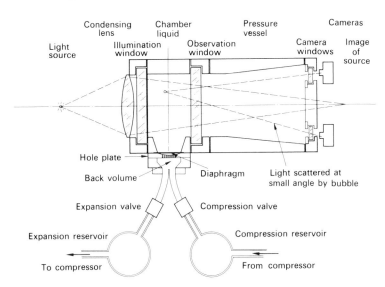

FIGURE F.3    The essential features of a bubble chamber. (From Bullock, 1970.)

from usually three different positions. The photographs are subsequently analysed to extract the required data on the particles which produced the tracks. The essential features of a bubble chamber are shown in Fig. F.3.

The heat track produced by a charged particle traversing a bubble chamber cools down in less than $10^{-6}$ s. As it takes several milliseconds to set the bubble chamber in operation, it is not possible to use counters to trigger a bubble chamber so as to select only particular

FIGURE F.4   Photograph of the BNL 80 in liquid hydrogen bubble chamber. The chamber, 80 in long by 27 in wide by 26 in deep is located in the centre of the magnet structure seen in the photograph. Pictures of tracks are taken from the left side where the cameras are visible. (From Shutt, 1967.)

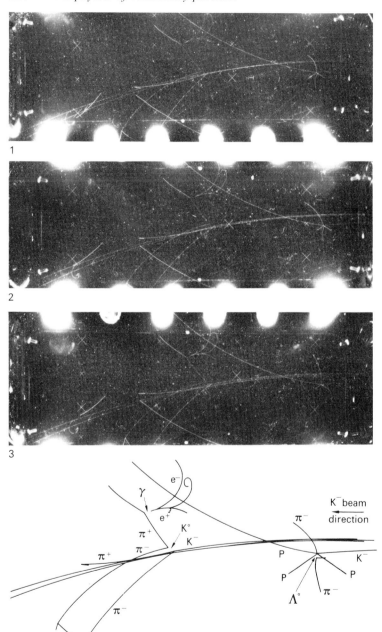

FIGURE F.5    Stereo triplet from the University College London/Rutherford Laboratory bubble chamber and interpretation of some of the events. (From Bullock, 1970.)

kinds of events. It is necessary for the liquid in the bubble chamber to be brought into a superheated state immediately before the ejection of particles from an accelerator, and to photograph the bubble chamber each time, and later examine the photographs, often numbering 100 000 to 1 000 000 for a given experiment. Because of the magnitude of the task of scanning bubble chamber photographs, considerable effort has gone into making it as automatic as possible with the help of computers (Alston *et al.*, 1967; Hough, 1967).

A variety of liquids have been used successfully in bubble chambers, ranging from liquid hydrogen as the least dense to xenon as the most dense. Early bubble chambers were small, with volumes of up to a few litres; but large bubble chambers with useful volumes of more than 20 000 litres have been built in order to increase the rate of observation of rare events, and to provide longer tracks so as to improve the accuracy of measurement.

The sensitive volume of a bubble chamber is usually embedded in a high magnetic field, so that the momentum of a particle can be determined from the radius of curvature of its track.

Some idea of the complexity of a bubble chamber can be gained from Fig. F.4. An example of bubble chamber photographs is shown in Fig. F.5. The tracks are curved because of the applied magnetic field.

A very readable introductory account of bubble chambers and their use is given by Thorndike in the initial and final chapters of Shutt (1967). For further accounts see Kenyon (1972), Bullock (1970) and Henderson (1970). Detailed information is available in Shutt (1967).

## F6  Spark chambers

There are several types of spark chamber. We first consider the track sampling spark chamber, also called the narrow gap spark chamber, which consists of a series of parallel electrodes with gaps of approximately 1 cm width, as shown in Fig. F.6. Between neighbouring electrodes an electric field is applied which is high enough to cause a discharge wherever the gas has been ionized, but not in the absence of ionization. Thus the track of a charged particle through the spark chamber is marked out by a series of sparks. The electric field is applied for only a short time, about 1 microsecond, and can be applied fast enough so that ionization formed by particles that have just passed through the chamber can cause sparks. The spark chamber has the advantage that it can be triggered into sensitivity using signals from particle counters. The simplest case, when the chamber is fired whenever a charged particle passes through it, is shown in Fig. F.6. In the narrow gap spark chamber, although each spark shows some tendency to follow the track of the ionizing particle, one essentially obtains samples of the trajectory at a series of discrete places.

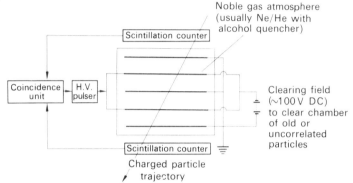

FIGURE F.6   Principles of triggered operation of a multiplate narrow gap spark chamber. (From Loebinger, 1970.)

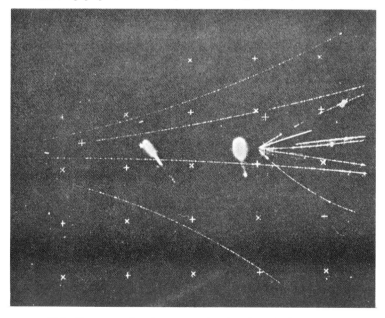

FIGURE F.7   Streamer chamber event showing four background tracks, a five-prong vertex and a two-prong decay (Bulos, F., A. Odian, F. Villa and D. Yount, 1967, SLAC Report No. 74.)

In wide gap spark chambers, the gap spacing is about 30–40 cm, and the sparks follow the trajectory of the ionizing particle up to angles of 45° to the applied field.

In the streamer chamber, the discharge is not allowed to fully develop, and only small visible streamers are left to indicate the path of the ionizing particle. The electrodes are transparent and the electric field is applied in a direction perpendicular to the particle trajectory. Figure F.7 shows an example of a streamer chamber photograph.

The first spark chambers used stereoscopic photography for recording spark positions. However, several automatic electronic techniques for providing direct digital read out of spark positions have been introduced. In some cases, these techniques allow data from spark chambers to be fed into computers for analysis. For further details of automatic spark chambers, also called filmless spark chambers, see Loebinger (1970) and Charpak (1970).

Introductory accounts of spark chambers are given by O'Neill (1962). A useful account is given by Loebinger (1970). For detailed information see Shutt (1967).

### References

ALSTON, M., J. V. FRANCK and L. T. KERTH, 'Conventional and Semi-automatic Data Processing and Interpretation', Vol. II of Shutt (1967), p. 52.

BIRKS, J. B., *The Theory and Practice of Scintillation Counting*, 1964. Pergamon Press, Oxford.

BLEWETT, M. H., 'Characteristics of typical accelerators', *Ann. Rev. Nucl. Sci.* **17** (1967) 427.

BLEWETT, M. H., 'Future prospects for high-energy accelerators', *Proc. Lund Int. Conf. on Elementary Particles*, 1969, p. 111.

BULLOCK, F. W., 'Bubble chambers', *Sci. Prog., Oxf.* **58** (1970) 301.

CHARPAK, G., 'Evolution of the automatic spark chambers', *Ann. Rev. Nucl. Sci.* **20** (1970) 195.

COURANT, E. D., 'Accelerators for high intensities and high energies', *Ann. Rev. Nucl. Sci.* **18** (1968) 435.

GIBSON, W. M., 'Counter experiments on elementary particles', *Sci. Prog., Oxf.* **58** (1970) 201.

GLASER, D. A., 'The bubble chamber', *Sci. Amer.*, February 1955. (Also available as reprint 214, Freeman, San Francisco.)

GOUIRAN, R., *Particles and Accelerators*, 1967, McGraw-Hill, New York.

HENDERSON, C., *Cloud and Bubble Chambers*, 1970. Methuen, London.

HOUGH, P. V. C., 'Fast precision digitizers on-line to computers for measurement and scanning', Vol. II of Shutt (1967), p. 141.

HUTCHINSON, G. W., 'Cerenkov detectors', *Prog. Nucl. Phys.* **8** (1960) 195.

ISR STAFF., 'First observation of colliding beam events in the CERN intersecting storage rings', *Phys. Lett.* **34B** (1971) 425. See also *Physics Today*, April 1971, p. 17.

KENYON, I. R., 'Bubble chamber film analysis', *Contemp. Phys.* **13** (1972) 75.

LOEBINGER, F. K., 'Spark chambers', *Sci. Prog., Oxf.* **58** (1970) 459.

LIVINGSTON, M. S. and BLEWETT, J. P., *Particle Accelerators*, 1962. McGraw-Hill, New York.

O'NEILL, G. K., 'The spark chamber', *Sci. Amer.*, August 1962. (Also available as reprint 282, Freeman, San Francisco); 'Particle storage rings', *Sci. Amer.* November 1966, p. 107.

POWELL, C. F., P. H. FOWLER and D. H. PERKINS, *The Study of Elementary Particles by the Photographic Method*, 1959. Pergamon, New York.

ROSENBLATT, J., *Particle Acceleration*, 1968. Methuen, London.

SEGRÈ, E., *Nuclei and Particles*, 1964. Benjamin, New York.

SHUTT, R. P. (editor), *Bubble and Spark Chambers*, 2 volumes, 1967. Academic Press, New York. In particular Thorndike, A. M., 'Introduction', 'Summary and future outlook'.

WILSON, R., 'Particle accelerators', *Sci. Amer.* March 1958. (Also available as reprint 251, Freeman, San Francisco.)

# List of particles

TABLE G.1   Properties of Particles. A summary of information from *Review of Particle Properties*, Particle Data Group, *Rev. Mod. Phys.* **43** (1971) No. 2, Part II

| Photon | | | |
|---|---|---|---|
| $\gamma$ | $J^P = 1^-$   zero mass | stable | |

**Leptons**

| Name | $J$ | Mass, $M$ (MeV) | Mean life (s) | Partial decay modes | |
|---|---|---|---|---|---|
| | | | | Mode | Fraction |
| $\nu_e$ | $\frac{1}{2}$ | $0 \, (<60\text{eV})$ | stable | | |
| $\nu_\mu$ | $\frac{1}{2}$ | $0 \, (<1\cdot6)$ | stable | | |
| $e$ | $\frac{1}{2}$ | $0\cdot5110041$ $\pm 0\cdot0000016$ | stable $(>2 \times 10^{21}$ years$)$ | | |
| $\mu$ | $\frac{1}{2}$ | $105\cdot6599$ $\pm 0\cdot0014$ | $2\cdot1983 \times 10^{-6}$ $\pm 0\cdot0008$ | $e\nu\bar{\nu}$ $e\gamma\gamma$ $3e$ $e\gamma$ | $100\%$ $<1\cdot6 \times 10^{-5}$ $<1\cdot3 \times 10^{-7}$ $<2\cdot2 \times 10^{-8}$ |

**HADRONS**

**Mesons**

| Name | $I^G$, $J^P$, $C$ (of neutrals) | Mass, $M$ (MeV) | Width, $\Gamma$ (MeV) | Partial decay modes | |
|---|---|---|---|---|---|
| | | | | Mode | Fraction (%) |

*Mesons with $Y = 0$. Symbol $\pi$ (with $I = 1$)*
*$\eta$ (with $I = 0$)*

| Name | $I^G$, $J^P$, $C$ (of neutrals) | Mass, $M$ (MeV) | Width, $\Gamma$ (MeV) | Partial decay modes | |
|---|---|---|---|---|---|
| $\pi^\pm$ | $1^-$, $0^-$ | $139\cdot576$ $\pm 0\cdot011$ | $0\cdot0$ Mean life $=$ $(2\cdot6024 \pm 0\cdot0024)$ $\times 10^{-8}$ s | $\mu\nu$ $e\nu$ $\mu\nu\gamma$ $\pi^0 e\nu$ $e\nu\gamma$ $e\nu e^+ e^-$ | $100$ $(1\cdot24 \pm 0\cdot03)10^{-2}$ $(1\cdot24 \pm 0\cdot25)10^{-2}$ $(1\cdot02 \pm 0\cdot07)10^{-6}$ $(3\cdot0 \pm 0\cdot5)10^{-6}$ $<3\cdot4 \times 10^{-6}$ |

| Name | $I^G$, $J^P$, $C$ (of neutrals) | Mass, $M$ (MeV) | Width, $\Gamma$ (MeV) | Partial decay modes | |
|---|---|---|---|---|---|
| | | | | Mode | Fraction (%) |
| $\pi^0$ | $1^-$, $0^-$, $+$ | 134·972 $\pm 0·012$ | 7·2 eV $\pm 1·2$ eV Mean life = $(0·84 \pm 0·10)$ $\times 10^{-16}$ s | $\gamma\gamma$ $\gamma e^+ e^-$ $\gamma\gamma\gamma$ $e^+ e^- e^+ e^-$ | $(98·84 \pm 0·04)$ $(1·16 \pm 0·04)$ $< 0·0005$ $0·00347$ |
| $\eta(549)$ | $0^+$, $0^-$, $+$ | 548·8 $\pm 0·6$ | 2·63 keV $\pm 0·59$ keV | $\gamma\gamma$ $\pi^0\gamma\gamma$ $3\pi^0$ | $38·6 \pm 1·1$ ⎫ Neutral $3·3 \pm 1·1$ ⎬ decays $30·3 \pm 1·1$ ⎭ 72·2% |
| | | | | $\pi^+\pi^-\pi^0$ $\pi^+\pi^-\gamma$ $\pi^0 e^+ e^-$ $\pi^+\pi^- e^+ e^-$ $\pi^+\pi^-\pi^0\gamma$ $\pi^+\pi^-\gamma\gamma$ $\mu^+\mu^-$ $\mu^+\mu^-\pi^0$ | $23·1 \pm 1·0$ $4·7 \pm 0·2$ $< 0·03$ $0·1 \pm 0·1$ ⎬ Charged $< 0·2$ ⎬ decays $< 0·2$ ⎬ 27·8% $(2 \pm 1)10^{-3}$ $< 0·05$ |
| $\eta_{0^+}$ (700–1000) or $\varepsilon$ | $0^+$, $0^+$, $+$ | $\gtrsim 750$ | $> > 100$ | $\pi\pi$ | 100 |
| $\rho(765)$ | $1^+$, $1^-$, $-$ | 765 $\pm 10$ | 125 $\pm 20$ | $\pi\pi$ $e^+ e^-$ $\mu^+\mu^-$ | $\approx 100$ $0·0060 \pm 0·0008$ $0·0067 \pm 0·0012$ |
| $\omega(784)$ | $0^-$, $1^-$, $-$ | 783·9 $\pm 0·3$ | 11·4 $\pm 0·9$ | $\pi^+\pi^-\pi^0$ $\pi^+\pi^-$ $\pi^0\gamma$ $e^+ e^-$ | $89·8 \pm 4·0$ $0·93 \pm 0·25$ $9·3 \pm 1·2$ $0·0066 \pm 0·0017$ |
| $\eta'(958)$ or $X^0$ | $0^+$, $-$, $+$ $J^P = 0^-$ or $2^-$ | 957·5 $\pm 0·8$ | $< 4$ | $\eta\pi\pi$ $\pi^+\pi^-\gamma$ (mainly $\rho^0\gamma$) $\gamma\gamma$ | $64·0 \pm 5·0$ $29·4 \pm 2·7$ $6·6 \pm 3·7$ |
| $\delta(962)$ | $\geqslant 1$ , , | 962 $\pm 5$ | $< 5$ | | ⎫ Interpretation of |
| $\pi_N(975)$ | $1^-$, $0^+$, $+$ | 975 $\pm 10$ | 58 $\pm 11$ | $\eta\pi$ possibly seen | ⎬ these three is |
| $\pi_N(1016)$ | $1^-$, $0^+$, $+$ | 1016 $\pm 10$ | $\approx 25$ | $K^\pm K^0$  only mode seen $\eta\pi$ | not clear $< 80$ ⎭ |
| $\phi(1019)$ | $0^-$, $1^-$, $-$ | 1019·5 $\pm 0·6$ | 4·0 $\pm 0·3$ | $K^+ K^-$ $K_L K_S$ $\pi^+\pi^-\pi^0$ $e^+ e^-$ $\mu^+\mu^-$ | $46·4 \pm 2·8$ $35·4 \pm 4·0$ $18·2 \pm 5·4$ $0·035 \pm 0·003$ $0·023 \pm 0·005$ |
| $\eta_{0^+}$ (1060) | $0^+$, $0^+$, $+$ | 1070 $\pm 30$ | 150–300 | $\pi\pi$ $KK$ | $< 65$ $> 35$ |
| $A1$ (1070) | $1^-$, $1^+$, $+$ | 1070 $\pm 20$ | 50–200 | $3\pi$ $KK$ | $\approx 100$ $< 0·25$ |

| Name | $I^G, J^P,$ $C$ (of neutrals) | Mass, $M$ (MeV) | Width, $\Gamma$ (MeV) | Partial decay modes | |
|------|------|------|------|------|------|
| | | | | Mode | Fraction (%) |
| $B(1235)$ | $1^+, 1^+, -$ | 1233 $\pm 10$ | 100 $\pm 20$ | $\omega\pi$ $\pi\pi$ $K\bar{K}$ | $\approx 100$ $< 30$ $< 2$ |
| $f(1260)$ | $0^+, 2^+, +$ | 1269 $\pm 10$ | 154 $\pm 25$ | $\pi\pi$ $2\pi^+2\pi^-$ $K\bar{K}$ | $\approx 80$ $7 \pm 2$ $\approx 5$ |
| $D(1285)$ | $0^+, A, +$ | 1286 $\pm 4$ | 33 $\pm 4$ | $K\bar{K}\pi$ [mainly $\pi_N(1016)$] $\pi\pi\eta$ $\pi_N(975)\pi$ $\pi\pi\rho$ | Seen Possibly large Possibly seen Not seen |
| $A2$ | $1^-, 2^+, +$ | $\approx 1300$ | | $\rho\pi$ $K\bar{K}$ $\eta\pi$ $\eta'(958)\pi$ | |
| $E(1422)$ | $0^+, 0^-, +$ | 1422 $\pm 4$ | 69 $\pm 8$ | $K^*\bar{K}+\bar{K}^*K$ $\pi_N(1016)\pi$ $\pi\pi\eta$ $\pi\pi\rho$ | $50 \pm 10$ $50 \pm 10$ $< 60$ Not seen |
| $f'(1514)$ | $0^+, 2^+, +$ | 1514 $\pm 5$ | 73 $\pm 23$ | $K\bar{K}$ $K^*\bar{K}+\bar{K}^*K$ $\pi\pi$ $\eta\pi\pi$ $\eta\eta$ | $72 \pm 12$ $10 \pm 10$ $< 14$ $18 \pm 10$ $< 40$ |
| $\pi/\rho$ (1540) or $F_1$ | $1, A$ | 1540 $\pm 5$ | 40 $\pm 15$ | $K^*\bar{K}+\bar{K}^*K$ | |
| $\pi_A(1640)$ | $1^-, A, +$ | 1640 $\pm 10$ | 50–200 | $f\pi$ $3\pi$ $\omega\pi\pi$ | Dominant Possibly observed |
| $\phi_N$ (1650) | $0^-, N, -$ | 1664 $\pm 13$ | 141 $\pm 17$ | $\rho\pi$ $3\pi$ $5\pi$ | Dominant Possibly observed $10 \pm 10$ |
| $\rho_N(1660)$ or $g$ | $1^+, N, -$ | 1660 $\pm 20$ | $\lesssim 200$ | $2\pi$ $K\bar{K}$ | Dominant $7 \pm 3$ |
| $\rho(1710)$ | $1^+, \ , -$ | 1712 $\pm 10$ | 125 $\pm 25$ | $4\pi$ $\pi^\pm A_2^0$ $\pi^\pm\omega$ $\rho^\pm\rho^0$ $\pi^\pm\phi$ $\pi^\pm 2\pi^+2\pi^-\pi^0$ $\pi\pi\rho$ | |

$\rho_N(1660)$ and $\rho(1710)$ may not be different resonances

| Name | $I^G, J^P$ | Mass, $M$ (MeV) | Width, $\Gamma$ (MeV) | Partial decay modes | |
|------|-----------|-----------------|----------------------|-------|--------|
| | | | | Mode | Fraction (%) |

*Mesons with $Y = \pm 1$. Symbol K*

| Name | $I^G, J^P$ | Mass, $M$ (MeV) | Width, $\Gamma$ (MeV) | Mode | Fraction (%) |
|------|-----------|-----------------|----------------------|-------|--------|
| $K^\pm$ | $\frac{1}{2}, 0^-$ | $493\cdot84$ $\pm 0\cdot11$ | Mean life = $(1\cdot2371$ $\pm 0\cdot0026)$ $\times 10^{-8}$ s | $\mu v$ | $63\cdot77 \pm 0\cdot28$ |
| | | | | $\pi\pi^0$ | $20\cdot92 \pm 0\cdot29$ |
| | | | | $\pi\pi^-\pi^+$ | $5\cdot58 \pm 0\cdot03$ |
| | | | | $\pi\pi^0\pi^0$ | $1\cdot68 \pm 0\cdot04$ |
| | | | | $\mu\pi^0 v$ | $3\cdot20 \pm 0\cdot11$ |
| | | | | $e\pi^0 v$ | $4\cdot86 \pm 0\cdot07$ |
| | | | | $\pi\pi^\mp e^\pm v$ | $(3\cdot3 \pm 0\cdot3)10^{-3}$ |
| | | | | $\pi\pi^\pm e^\mp v$ | $(<7)\quad 10^{-5}$ |
| | | | | $\pi\pi^\mp \mu^\pm v$ | $(0\cdot9 \pm 0\cdot4)10^{-3}$ |
| | | | | $\pi\pi^\pm \mu^\mp v$ | $(<3)\quad 10^{-4}$ |
| | | | | $ev$ | $(1\cdot30 \pm 0\cdot18)10^{-3}$ |
| | | | | $\pi\pi^0\gamma$ | $(<1\cdot9)\quad 10^{-2}$ |
| | | | | $\pi\pi^+\pi^-\gamma$ | $(10 \pm 4)\quad 10^{-3}$ |
| | | | | $\pi ev\gamma$ | $(6 \pm 4)\quad 10^{-2}$ |
| | | | | $\pi e^+ e^-$ | $(<0\cdot4)\quad 10^{-4}$ |
| | | | | $\pi\mu^+\mu^-$ | $(<2\cdot4)\quad 10^{-4}$ |
| | | | | $\pi\gamma\gamma$ | $(<0\cdot4)\quad 10^{-2}$ |
| | | | | $\pi v\bar{v}$ | $(<1\cdot2)\quad 10^{-4}$ |
| | | | | $\pi\gamma$ | $(<4)\quad 10^{-4}$ |
| $K^0$ | $\frac{1}{2}, 0^-$ | $497\cdot79 \pm 0\cdot15$ | $50\% \ K_{\text{Short}}$ | $50\% \ K_{\text{Long}}$ | |
| $K^0_S$ | $\frac{1}{2}, 0^-$ | | Mean life = $(0\cdot862$ $\pm 0\cdot006)$ $\times 10^{-10}$ | $\pi^+\pi^-$ | $68\cdot7 \pm 0\cdot5$ |
| | | | | $\pi^0\pi^0$ | $31\cdot3 \pm 0\cdot5$ |
| | | | | $\mu^+\mu^-$ | $(<0\cdot7)10^{-3}$ |
| | | | | $e^+e^-$ | $<0\cdot035$ |
| | | | | $\pi^+\pi^-\gamma$ | $0\cdot23 \pm 0\cdot08$ |
| $K^0_L$ | $\frac{1}{2}, 0^-$ | | Mean life = $(5\cdot172$ $\pm 0\cdot043)$ $\times 10^{-8}$ | $\pi^0\pi^0\pi^0$ | $21\cdot4 \pm 0\cdot7$ |
| | | | | $\pi^+\pi^-\pi^0$ | $12\cdot6 \pm 0\cdot3$ |
| | | | | $\pi\mu v$ | $26\cdot8 \pm 0\cdot6$ |
| | | | | $\pi ev$ | $38\cdot9 \pm 0\cdot6$ |
| | | | | $\pi^+\pi^-$ | $0\cdot157 \pm 0\cdot005$ |
| | | | | $\pi^0\pi^0$ | $0\cdot094 \pm 0\cdot019$ |
| | | | | $\pi^+\pi^-\gamma$ | $<0\cdot04$ |
| | | | | $\gamma\gamma$ | $(5\cdot6 \pm 0\cdot5)10^{-2}$ |
| | | | | $e\mu$ | $(<1\cdot6)\quad 10^{-7}$ |
| | | | | $\mu^+\mu^-$ | $(<1\cdot9)\quad 10^{-7}$ |
| | | | | $e^+e^-$ | $(<1\cdot6)\quad 10^{-7}$ |
| $K^*(892)$ | $\frac{1}{2}, 1^-$ | $892\cdot6$ $\pm 0\cdot5$ | $50\cdot3$ $\pm 1\cdot1$ | $K\pi$ | $\approx 100$ |
| | | | | $K\pi\pi$ | $0\cdot2$ |
| $K_A$ (1240) or $C$ | $\frac{1}{2}, 1^+$ | $1242$ $\pm 10$ | $127$ $\pm 25$ | $K\pi\pi$ | |
| $K_N$ (1420) or $K^{**}$ | $\frac{1}{2}, 2^+$ | $1408$ $\pm 10$ | $107$ $\pm 15$ | $K\pi$ | $56\cdot9 \pm 4\cdot0$ |
| | | | | $K^*\pi$ | $27\cdot4 \pm 3\cdot2$ |
| | | | | $K\rho$ | $9\cdot2 \pm 3\cdot5$ |
| | | | | $K\omega$ | $4\cdot5 \pm 1\cdot8$ |
| | | | | $K\eta$ | $2\cdot0 \pm 1\cdot8$ |
| $L(1770)$ | $\frac{1}{2}, A$ | $1770$ $\pm 10$ | $50$–$140$ | $K\pi\pi$ | Dominant |
| | | | | $K\pi\pi\pi$ | Possibly seen |

**Baryons**

| Name | $I^G, J^P$ | Mass, $M$ (MeV) | Width, $\Gamma$ (MeV) | Partial decay modes | |
|---|---|---|---|---|---|
| | | | | Mode | Fraction (%) |

*Baryons with $Y = 1$, $I = \frac{1}{2}$. Symbol N*

| Name | $I^G, J^P$ | Mass, $M$ (MeV) | Width, $\Gamma$ (MeV) | Mode | Fraction (%) |
|---|---|---|---|---|---|
| $p$ | $\frac{1}{2}, \frac{1}{2}^+$ | 938·2592 ±0·0052 | stable Mean life $> 2 \times 10^{28}$ years | | |
| $n$ | $\frac{1}{2}, \frac{1}{2}^+$ | 939·5527 ±0·0052 | Mean life = $(0·932 ±0·014) \times 10^3$ s | $pe^-\nu$ | 100 |
| $N'(1470)$ | $\frac{1}{2}, \frac{1}{2}^+$ | 1435–1505 | 165–400 | $N\pi$ $N\pi\pi$ | 60 40 |
| $N'(1520)$ | $\frac{1}{2}, \frac{3}{2}^-$ | 1510–1540 | 105–150 | $N\pi$ $N\pi\pi$ $N\eta$ | 50 50 $\sim0·6$ |
| $N'(1535)$ | $\frac{1}{2}, \frac{1}{2}^-$ | 1500–1600 | 50–160 | $N\pi$ $N\eta$ $N\pi\pi$ | 35 55 $\sim10$ |
| $N(1670)$ | $\frac{1}{2}, \frac{5}{2}^-$ | 1655–1680 | 105–175 | $N\pi$ $N\pi\pi$ $\Lambda K$ $N\eta$ | 40 60 $<0·3$ $<1$ |
| $N(1688)$ | $\frac{1}{2}, \frac{5}{2}^+$ | 1680–1692 | 105–180 | $N\pi$ $N\pi\pi$ $\Lambda K$ $N\eta$ | 60 40 $<0·2$ $<0·5$ |
| $N''(1700)$ | $\frac{1}{2}, \frac{1}{2}^-$ | 1665–1765 | 100–400 | $N\pi$ $\Lambda K$ $N\eta$ | 65 5 |
| $N''(1780)$ | $\frac{1}{2}, \frac{1}{2}^+$ | 1650–1860 | 50–450 | $N\pi$ $\Lambda K$ $N\eta$ | 30 $\sim7$ $\sim10$ |
| $N(1860)$ | $\frac{1}{2}, \frac{3}{2}^+$ | 1770–1900 | 180–330 | $N\pi$ $N\pi\pi$ $\Lambda K$ $N\eta$ | 25 $\sim5$ $\sim4$ |
| $N(2190)$ | $\frac{1}{2}, \frac{7}{2}^-$ | 2000–2260 | 270–325 | $N\pi$ $N\pi\pi$ | 25 |
| $N(2220)$ | $\frac{1}{2}, \frac{9}{2}^+$ | 2200–2245 | 260–330 | $N\pi$ $N\pi\pi$ | 15 |
| $N(2650)$ | $\frac{1}{2}, ?^-$ | 2650 | 360 | $N\pi$ $N\pi\pi$ | |
| $N(3030)$ | $\frac{1}{2}, ?$ | 3030 | 400 | $N\pi$ $N\pi\pi$ | |

| Name | $I^G$, $J^P$ | Mass, $M$ (MeV) | Width, $\Gamma$ (MeV) | Partial decay modes | |
| --- | --- | --- | --- | --- | --- |
| | | | | Mode | Fraction (%) |

*Baryons with $Y = 1$, $I = \frac{3}{2}$. Symbol $\Delta$*

| Name | $I^G$, $J^P$ | Mass, $M$ (MeV) | Width, $\Gamma$ (MeV) | Mode | Fraction (%) |
| --- | --- | --- | --- | --- | --- |
| $\Delta(1236)$ | $\frac{3}{2}, \frac{3}{2}^+$ | 1230–1236 | 110–122 | $N\pi$ | 99·4 |
| | | | | $N\pi^+\pi^-$ | 0 |
| | | | | $N\gamma$ | $\sim 0·6$ |
| $\Delta(1650)$ | $\frac{3}{2}, \frac{1}{2}^-$ | 1615–1695 | 130–200 | $N\pi$ | 28 |
| | | | | $N\pi\pi$ | 72 |
| $\Delta(1670)$ | $\frac{3}{2}, \frac{3}{2}^-$ | 1650–1720 | 175–300 | $N\pi$ | 15 |
| | | | | $N\pi\pi$ | |
| $\Delta(1890)$ | $\frac{3}{2}, \frac{5}{2}^+$ | 1840–1920 | 135–350 | $N\pi$ | 17 |
| | | | | $N\pi\pi$ | |
| $\Delta(1910)$ | $\frac{3}{2}, \frac{1}{2}^+$ | 1780–1935 | 230–420 | $N\pi$ | 25 |
| | | | | $N\pi\pi$ | |
| $\Delta(1950)$ | $\frac{3}{2}, \frac{7}{2}^+$ | 1930–1980 | 140–220 | $N\pi$ | 45 |
| | | | | $\Delta(1236)\pi$ | $\approx 50$ |
| | | | | $\Sigma K$ | $\sim 2$ |
| | | | | $\Sigma(1385)K$ | 1·4 |
| $\Delta(2420)$ | $\frac{3}{2}, \frac{11}{2}^+$ | 2320–2450 | 270–350 | $N\pi$ | 11 |
| | | | | $N\pi\pi$ | $> 20$ |
| $\Delta(2850)$ | $\frac{3}{2}, ?^+$ | 2850 | 400 | $N\pi$ | |
| | | | | $N\pi\pi$ | |
| $\Delta(3230)$ | $\frac{3}{2}, ?$ | 3230 | 440 | $N\pi$ | |
| | | | | $N\pi\pi$ | |

*Baryons with $Y = 0$, $I = 0$. Symbol $\Lambda$*

| Name | $I^G$, $J^P$ | Mass, $M$ (MeV) | Width, $\Gamma$ (MeV) | Mode | Fraction (%) |
| --- | --- | --- | --- | --- | --- |
| $\Lambda$ | $0, \frac{1}{2}^+$ | $1115·59 \pm 0·06$ | Mean life $= (2·517 \pm 0·024) \times 10^{-10}$ s | $p\pi^-$ | $\left.\begin{array}{c}64·0 \\ 36·0\end{array}\right\} \pm 0·7$ |
| | | | | $n\pi^0$ | |
| | | | | $pe\nu$ | $0·080 \pm 0·006$ |
| | | | | $p\mu\nu$ | $(1·35 \pm 0·60)10^{-2}$ |
| $\Lambda(1405)$ | $0, \frac{1}{2}^-$ | $1405 \pm 5$ | $40 \pm 10$ | $\Sigma\pi$ | 100 |
| $\Lambda'(1520)$ | $0, \frac{3}{2}^-$ | $1518 \pm 2$ | $16 \pm 2$ | $N\overline{K}$ | $46 \pm 1$ |
| | | | | $\Sigma\pi$ | $41 \pm 1$ |
| | | | | $\Lambda\pi\pi$ | $9·6 \pm 0·7$ |
| | | | | $\Sigma\pi\pi$ | $1·0 \pm 0·1$ |
| $\Lambda'(1670)$ | $0, \frac{1}{2}^-$ | 1670 | 15–38 | $N\overline{K}$ | $\sim 20$ |
| | | | | $\Lambda\eta$ | $\sim 35$ |
| | | | | $\Sigma\pi$ | $\sim 45$ |
| $\Lambda''(1690)$ | $0, \frac{3}{2}^-$ | 1690 | 27–85 | $N\overline{K}$ | $\sim 30$ |
| | | | | $\Sigma\pi$ | $\sim 40$ |
| | | | | $\Lambda\pi\pi$ | $\sim 20$ |
| | | | | $\Sigma\pi\pi$ | $\sim 10$ |
| $\Lambda(1815)$ | $0, \frac{5}{2}^+$ | $1820 \pm 5$ | 64 to 100 | $N\overline{K}$ | 62 |
| | | | | $\Sigma\pi$ | 11 |
| | | | | $\Sigma(1385)\pi$ | 17 |
| $\Lambda(1830)$ | $0, \frac{5}{2}^-$ | 1835 | 74–150 | $N\overline{K}$ | $\sim 10$ |
| | | | | $\Sigma\pi$ | $\sim 30$ |

| Name | $I^G$, $J^P$ | Mass, $M$ (MeV) | Width, $\Gamma$ (MeV) | Partial decay modes | |
| --- | --- | --- | --- | --- | --- |
| | | | | Mode | Fraction (%) |
| $\Lambda(2100)$ | $0, \frac{7}{2}^-$ | 2100 | 60–140 | $N\bar{K}$ | 25 |
| | | | | $\Sigma\pi$ | $\sim 5$ |
| | | | | $\Lambda\eta$ | $< 3$ |
| | | | | $\Xi K$ | |
| | | | | $\Lambda\omega$ | $< 10$ |
| $\Lambda(2350)$ | $0,\ ?$ | 2350 | 140–324 | $N\bar{K}$ | |

*Baryons with $Y = 0$, $I = 1$. Symbol $\Sigma$*

| Name | $I^G$, $J^P$ | Mass, $M$ (MeV) | Width, $\Gamma$ (MeV) | Partial decay modes | |
| --- | --- | --- | --- | --- | --- |
| $\Sigma^+$ | $1, \frac{1}{2}^+$ | $1189.42$ $\pm 0.11$ | Mean life = $(0.800$ $\pm 0.006)$ $\times 10^{-10}$ | $p\pi^0$ | $51.7$ |
| | | | | $n\pi^+$ | $48.3 \pm 0.8$ |
| | | | | $p\gamma$ | $0.124 \pm 0.018$ |
| | | | | $n\pi^+\gamma$ | $(1.30 \pm 0.24)10^{-2}$ |
| | | | | $\Lambda e^+ v$ | $(2.02 \pm 0.47)10^{-3}$ |
| | | | | $n\mu^+ v$ | $(< 2.4)\ \ \ 10^{-3}$ |
| | | | | $n e^+ v$ | $(< 1.0)\ \ \ 10^{-3}$ |
| $\Sigma^0$ | $1, \frac{1}{2}^+$ | $1192.51$ $\pm 0.10$ | Mean life = $< 1.0$ $\times 10^{-14}$ s | $\Lambda\gamma$ | 100 |
| | | | | $\Lambda e^+ e^-$ | 0.545 |
| $\Sigma^-$ | $1, \frac{1}{2}^+$ | $1197.37$ $\pm 0.07$ | Mean life = $(1.489$ $\pm 0.022)$ $\times 10^{-10}$ s | $n\pi^-$ | 100 |
| | | | | $n e^- v$ | $0.109 \pm 0.005$ |
| | | | | $n\mu^- v$ | $0.045 \pm 0.004$ |
| | | | | $\Lambda e^- v$ | $(0.6 \pm 0.06)10^{-2}$ |
| | | | | $n\pi^-\gamma$ | $(1.0 \pm 0.2)\ 10^{-2}$ |
| $\Sigma(1385)$ | $1, \frac{3}{2}^+$ | $(+)1383 \pm 1$ $(-)1386 \pm 2$ | $(+)36 \pm 3$ $(-)36 \pm 6$ | $\Lambda\pi$ | $90 \pm 3$ |
| | | | | $\Sigma\pi$ | $10 \pm 3$ |
| $\Sigma(1670)$ | $1, \frac{3}{2}^-$ | 1670 | 50 | $N\bar{K}$ | $\sim 8$ |
| | | | | $\Sigma\pi$ | |
| | | | | $\Lambda\pi$ | |
| | | | | $\Sigma\pi\pi$ | |
| | | | | $\Lambda\pi\pi$ | |
| $\Sigma(1750)$ | $1, \frac{1}{2}^-$ | 1750 | 50–80 | $N\bar{K}$ | $\sim 15$ |
| | | | | $\Lambda\pi$ | seen |
| | | | | $\Sigma\eta$ | seen |
| $\Sigma(1765)$ | $1, \frac{5}{2}^-$ | $1765 \pm 5$ | $\sim 120$ | $N\bar{K}$ | $\sim 44$ |
| | | | | $\Lambda\pi$ | $\sim 15$ |
| | | | | $\Lambda(1520)\pi$ | $\sim 14$ |
| | | | | $\Sigma(1385)\pi$ | $\sim 13$ |
| | | | | $\Sigma\pi$ | $\sim 1$ |
| $\Sigma(1915)$ | $1, \frac{5}{2}^+$ | 1910 | 70 | $N\bar{K}$ | $\sim 11$ |
| | | | | $\Lambda\pi$ | |
| | | | | $\Sigma\pi$ | |
| $\Sigma(2030)$ | $1, \frac{7}{2}^+$ | 2030 | 100–170 | $N\bar{K}$ | 10–27 |
| | | | | $\Lambda\pi$ | 14–38 |
| | | | | $\Sigma\pi$ | 2–5 |
| | | | | $\Xi K$ | $< 2$ |

| Name | $I^G$, $J^P$ | Mass, $M$ (MeV) | Width, $\Gamma$ (MeV) | Partial decay modes | |
|---|---|---|---|---|---|
| | | | | Mode | Fraction (%) |
| $\Sigma(2250)$ | 1, ? | 2250 | 100–230 | $N\bar{K}$ $\Sigma\pi$ $\Lambda\pi$ | |
| $\Sigma(2455)$ | 1, ? | 2455 | ~120 | $N\bar{K}$ | |
| $\Sigma(2620)$ | 1, ? | 2620 | ~175 | $N\bar{K}$ | |

*Baryons with $Y = -1$, $I = \frac{1}{2}$. Symbol $\Xi$*

| Name | $I^G$, $J^P$ | Mass, $M$ (MeV) | Width, $\Gamma$ (MeV) | Mode | Fraction (%) |
|---|---|---|---|---|---|
| $\Xi^0$ | $\frac{1}{2}$, $\frac{1}{2}^+$ | 1314·7 ±0·7 | Mean life = $(3·03 ±0·18) \times 10^{-10}$ | $\Lambda\pi^0$ $p\pi^-$ $pe^-v$ $\Sigma^+e^-v$ $\Sigma^-e^+v$ $\Sigma^+\mu^-v$ $\Sigma^-\mu^+v$ $p\mu^-v$ | 100 <0·09 <0·13 <0·15 <0·15 <0·15 <0·15 <0·13 |
| $\Xi^-$ | $\frac{1}{2}$, $\frac{1}{2}^+$ | 1321·31 ±0·17 | Mean life = $(1·660 ±0·037) \times 10^{-10}$ s | $\Lambda\pi^-$ $\Lambda e^-v$ $\Sigma^0e^-v$ $\Lambda\mu^-v$ $\Sigma^0\mu^-v$ $n\pi^-$ $ne^-v$ | 100 $0·067 \pm 0·023$ <0·05 <0·13 <0·5 <0·11 <1·0 |
| $\Xi(1530)$ | $\frac{1}{2}$, $\frac{3}{2}^+$ | (0)1528·9 ±1·1 (−)1533·8 ±1·9 | 7·3 ± 1·7 | $\Xi\pi$ | 100 |
| $\Xi(1820)$ | $\frac{1}{2}$, ? | 1795–1870 | 12–99 | $\Lambda\bar{K}$ $\Xi\pi$ $\Xi(1530)\pi$ $\Sigma K$ | All four decay modes have been seen |
| $\Xi(1940)$ | $\frac{1}{2}$, ? | 1894–1961 | 42–140 | $\Xi\pi$ $\Xi(1530)\pi$ | |

*Baryons with $Y = -2$, $I = 0$. Symbol $\Omega$*

| Name | $I^G$, $J^P$ | Mass, $M$ (MeV) | Width, $\Gamma$ (MeV) | Mode | Fraction (%) |
|---|---|---|---|---|---|
| $\Omega^-$ | 0, $\frac{3}{2}^+$ | 1672·5 ± 0·5 | Mean life = $\left(1·3{+0·4 \atop -0·3}\right) \times 10^{-10}$ s | $\Xi^0\pi^-$ $\Xi^-\pi^0$ $\Lambda K^-$ | |

*Notes.* In spin and parity assignments for mesons, $N$ stands for $J^P = 0^+$, $1^-$, $2^+$, $3^-$, etc.
$A$ stands for $J^P = 0^-$, $1^+$, $2^-$, $3^+$, etc.

# Physical constants

Values are from the adjustment of fundamental physical constants by B. N. Taylor, W. H. Parker and D. N. Langenberg, *Rev. Mod. Phys.* **41** (1969) 375. The numbers in parentheses are the standard-deviation uncertainties in the last digits of the quoted value.

$$\text{The velocity of light, } c = 2\cdot997\ 9250(10) \times \begin{cases} 10^{10} \text{ cm s}^{-1} \\ 10^{8} \text{ m s}^{-1} \end{cases}$$

$$\text{Fine structure constant, } \alpha = \frac{e^2}{\hbar c} = 7\cdot297\ 351(11) \times 10^{-3}$$

$$\frac{1}{\alpha} = 137\cdot036\ 02(21)$$

$$\text{Electron charge, } e = 4\cdot803\ 250(21) \times 10^{-10} \text{ esu}$$
$$= 1\cdot602\ 1917(70) \times 10^{-19} \text{ coulomb}$$

$$\text{Planck's constant, } h = 6\cdot626\ 196(50) \times 10^{-27} \text{ erg s}$$
$$\hbar = \frac{h}{2\pi} = 1\cdot054\ 5919(80) \times 10^{-27} \text{ erg s}$$

$$\text{Electron rest mass, } M_e = 9\cdot109\ 558(54) \times 10^{-31} \text{ kg}$$
$$= 0\cdot511\ 0041(16) \text{ MeV}$$

$$(1 \text{ MeV} = 10^6 \text{ eV})$$

$$\text{Proton rest mass, } M_p = 1\cdot672\ 614(11) \times 10^{-27} \text{ kg}$$
$$= 938\cdot2592(52) \text{ MeV}$$

$$\text{Neutron rest mass, } M_n = 1\cdot674\ 920(11) \times 10^{-27} \text{ kg}$$
$$= 939\cdot5527(52) \text{ MeV}$$

$$\text{Bohr magneton, } \mu_{\text{Bohr}} = \frac{e\hbar}{2 M_e c}$$
$$= 9\cdot274\ 096(65) \times \begin{cases} 10^{-21} \text{ erg gauss}^{-1} \\ 10^{-24} \text{ J T}^{-1} \end{cases}$$
$$= 5\cdot788\ 381(18) \times 10^{-5} \text{ eV T}^{-1}$$

$(T = \text{tesla} = 10^4 \text{ gauss. J} = \text{joule})$

Nuclear magneton, $\mu_{\text{nuclear}} = \dfrac{eh}{2M_p c}$

$$= 5{\cdot}050\ 951(50) \times \begin{cases} 10^{-24}\ \text{erg gauss}^{-1} \\ 10^{-27}\ \text{J T}^{-1} \end{cases}$$

$$= 3{\cdot}152\ 526(21) \times 10^{-8}\ \text{eV T}^{-1}$$

$(1\ \text{MeV} = 1{\cdot}602\ 1917(70) \times 10^{-6}\ \text{erg})$

$1\text{b} = 1$ barn (cross-section) $= 10^{-24}\ \text{cm}^2$

$$1\text{mb} = 10^{-3}\text{b}$$

$$1\mu\text{b} = 10^{-6}\text{b}$$

$$1\text{nb} = 10^{-9}\text{b}$$

$$1\ \text{fermi} = 10^{-13}\ \text{cm}$$

$$1\ \text{fermi}^2 = 10\ \text{millibarns}$$

Momentum $p$ is usually expressed in units of $\text{GeV}/c$.

$$p(\text{GeV}/c) = \{[E(\text{GeV})]^2 + [M(\text{GeV})]^2\}^{\frac{1}{2}}$$

$(1\ \text{GeV} = 1\ \text{BeV} = 10^9\ \text{eV})$

# Answers to even-numbered exercises

## Chapter 1

2. The de Broglie wavelength is given by

$$\lambda = h/p$$

Using

$$p^2 = (E^2 - m^2 c^4)/c^2$$

and writing

$$E = T + mc^2$$
$$\lambda = hc[T(T + 2mc^2)]^{-\frac{1}{2}}$$

(a) $3.9 \times 10^{-8}$ cm
(b) $1.40 \times 10^{-10}$ cm
(c) $1.23 \times 10^{-12}$ cm

4. (a) 145 MeV,     (b) 0.079 MeV

## Chapter 2

2. In the centre-of-mass system, the total momentum

$$P_c = 0$$

and for pair production at threshold, the final state consists of two electrons and one positron all at rest with total energy

$$E_c = 3Mc^2$$

In the laboratory system, in which the initial electron is at rest,

$$E_L = h\nu + Mc^2$$
$$P_L = h\nu/c$$

Since $E^2 - c^2 P^2$ is a relativistic invariant,

$$(3Mc^2)^2 = (h\nu + Mc^2)^2 - (h\nu)^2$$
$$\therefore \ h\nu = 4Mc^2$$

4.

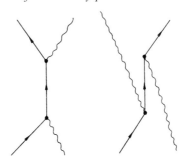

FIGURE I.1.

6.

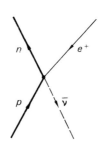

FIGURE I.2.

8. (a)
$$\pi^+ \rightarrow \mu^+ + \nu_\mu$$
$$\pi^- \rightarrow \mu^- + \bar{\nu}_\mu$$

$\nu_\mu$, $\bar{\nu}_\mu$ have zero charge.

$$Q = M_\pi c^2 - M_\mu c^2 = 33.9 \text{ MeV}$$

For initial pion stationary,

$$\mathbf{p}_\nu = -\mathbf{p}_\mu = \mathbf{p}$$
$$E_\mu + E_\nu = M_\pi c^2$$
$$c(p^2 + M_\mu^2 c^2)^{\frac{1}{2}} + cp = M_\pi c^2$$

Solving for $p$

$$p = c(M_\pi^2 - M_\mu^2)/2M_\pi$$

and

$$E_\mu = c^2(M_\pi^2 + M_\mu^2)/2M_\pi$$
$$T_\mu = E_\mu - M_\mu c^2$$
$$= c^2(M_\pi - M_\mu)^2/2M_\pi$$
$$= 4.12 \text{ MeV}$$

(b) $$\pi^0 \to \gamma + \gamma$$

$\gamma$ is uncharged. $Q = 135\cdot0$ MeV

(c) $$\mu^+ \to e^+ + v_e + \bar{v}_\mu$$
$$\mu^- \to e^- + \bar{v}_e + v_\mu$$

We distinguish between the neutrino associated with the muon and the neutrino associated with the electron, for we shall see in Section 34 that there are two different kinds of neutrino. $v_e$, $\bar{v}_e$, $v_\mu$, $\bar{v}_\mu$ have zero charge. $Q = 105\cdot2$ MeV.

10. In the laboratory system, at threshold the incident proton has a kinetic energy of $6M_p c^2$.

## Chapter 3

2. (a) $+1$,    (b) $+1$,    (c) $-1$

4. (a), (e) forbidden because they do not conserve baryon number; (c), (d), (g) forbidden because they do not conserve charge.

## Chapter 4

2. For $L=1$, $\psi(\mathbf{r})$ is antisymmetric with respect to interchange of the two nucleons.

(a) For antisymmetry of

$$\Psi = \psi(\mathbf{r})\psi_{\text{spin}}\psi_{\text{isospin}}$$

we require $\psi_{\text{spin}}\psi_{\text{isospin}}$ to be symmetric. There are 10 such states as shown in Table I1.

TABLE I.1

| Spin state | Isospin state | Number of states |
|---|---|---|
| Symmetric $S = 1$ 3 states | Symmetric $I = 1$ 3 states | $3 \times 3 = 9$ |
| Antisymmetric $S = 0$ 1 state | Antisymmetric $I = 0$ 1 state | 1 |
| | | Total 10 |

(b)  The 10 states are shown in Table I.2

TABLE I.2

|     |           | States |
| --- | --------- | ------ |
| $nn$ | $S = 1$   | 3      |
| $pp$ | $S = 1$   | 3      |
| $np$ | $S = 0, 1$ | 4     |
|     |           | Total  10 |

4. (a) $\frac{1}{2}, \frac{3}{2}$.    (b) since $I_3 = +\frac{3}{2}, I = \frac{3}{2}$

## Chapter 6

2. In the rest frame of $A$,

$$\mathbf{p}_B = -\mathbf{p}_C = \mathbf{p}$$

From conservation of energy

$$M_A c^2 = c(p^2 + M_B^2 c^2)^{\frac{1}{2}} + c(p^2 + M_C^2 c^2)^{\frac{1}{2}}$$
$$= E_B + \{E_B^2 + (M_C^2 - M_B^2)c^4\}^{\frac{1}{2}}$$

Solving for $E_B$

$$E_B = \frac{(M_A^2 + M_B^2 - M_C^2)c^2}{2M_A}$$
$$T_B = E_B - M_B c^2$$
$$= \frac{c^2}{2M_A} \{(M_A - M_B)^2 - M_C^2\}$$

4.  The processes which cannot occur through the strong interactions and the quantities they would not conserve are

      (a) strangeness,    (b) strangeness,
      (d) charge,        (e) energy

## Chapter 7

2. (b) Totally forbidden by charge conservation; (c), (g), (h) weak, do not conserve $P$, $S$, $I$ and $I_3$; (a), (f) electromagnetic, do not conserve $I$, but conserve $P$, $S$, $I_3$; (d), (e) strong, conserve $P$, $S$, $I$ and $I_3$.

## Chapter 8

2. Since the spin of the pion is 0, the distribution of neutrinos is isotropic in the rest frame of the pion. The distribution of neutrinos in the laboratory frame is then proportional to d cos $\theta'$/d cos $\theta$ given by equation (A.68) with $M = 0$,

$$\frac{1}{\gamma^2} \frac{1}{(1 - \beta \cos \theta)^2}$$

| 4. | Decay | Branching ratio |
|---|---|---|
| | $\Lambda^0 \to p + \pi^-$ | 64% |
| | $\to n + \pi^0$ | 36% |
| | $\to p + e^- + \bar{\nu}_e$ | $0.8 \times 10^{-3}$ |
| | $\to p + \mu^- + \bar{\nu}_\mu$ | $1.6 \times 10^{-4}$ |
| | $\to n + e^- + \bar{\nu}_e + \mu^+ + \nu_\mu$ | |
| | $\to n + e^+ + \nu_e + \mu^- + \bar{\nu}_\mu$ | |
| | $\to$ all of above $+ \gamma$ | |
| | $\to$ all of above $+ e^+ + e^-$ | |

Branching ratios are taken from *Review of Particle Properties*, 1972. (*Phys. Letters* **39B**, No. 1.)

## Chapter 9

2. Intensity is given by

$$I = I_0 \, e^{-t/\tau_{\text{lab}}}$$

where $\tau_{\text{lab}}$ is the mean life in the laboratory frame. Using equation (A.71),

$$I = I_0 \, e^{-\pi/\tau v \gamma}$$

where $\tau$ is the mean life in the rest frame.

$$\gamma = E/mc^2$$

For 10 GeV kaons,

$$\gamma \simeq 20$$
$$v \simeq c$$

For $x = 20 \times 10^2$ cm,

$$I = I_0 \, e^{-10^2/c\tau(cm)}$$

For $K_S^0$, $c\tau = 2{\cdot}59$ cm, and

$$I/I_0 \simeq 10^{-17}$$

For $K_L^0$, $c\tau = 1614$ cm, and the decay of $K_L^0$ is negligible. Ratio of $K_S^0$ to $K_L^0$ is $10^{-17}$

## Chapter 10

2. $E = Mc^2$, where $M$ is the invariant mass of the pion–proton system. Then

$$W^2 = c^2(P^2 + M^2 c^2) \tag{1}$$

where $W$ is the total energy and $P$ the total momentum in an arbitrary reference frame. In the laboratory frame

$$W = M_\pi c^2 + M_p c^2 + T_\pi \tag{2}$$

$$\begin{aligned} P = p_\pi &= [(T_\pi + M_\pi c^2)^2 - M_\pi^2 c^4]^{\frac{1}{2}}/c \\ &= [T_\pi(T_\pi + 2M_\pi c^2)]^{\frac{1}{2}}/c \end{aligned} \tag{3}$$

Substituting (2) and (3) into (1),

$$\begin{aligned} T_\pi &= \left[\frac{M^2 - (M_p + M_\pi)^2}{2M_p}\right]c^2 \\ &= \frac{(M + M_p + M_\pi)\{M - (M_p + M_\pi)\}c^2}{2M_p} \end{aligned} \tag{4}$$

Substituting (4) into (3)

$$p_\pi = \frac{c}{2M_p}[\{M^2 - (M_p + M_\pi)^2\}\{M^2 - (M_p - M_\pi)^2\}]^{\frac{1}{2}}$$

With

$M_\pi c^2 = 0{\cdot}1396$ GeV

$M_p c^2 = 0{\cdot}9383$ GeV

Equation (4) yields

$$T_\pi(\text{GeV}) = \frac{\{E(\text{GeV}) + 1{\cdot}0779\}\{E\,\text{GeV}) - 1{\cdot}0779\}}{1{\cdot}8766} \tag{5}$$

and equation (3) yields

$$p_\pi\,(\text{GeV}/c) = [T_\pi(\text{GeV})\{T(\text{GeV}) + 0{\cdot}2792\}]^{\frac{1}{2}} \tag{6}$$

See Table I.3.

TABLE I.3

| | $T_\pi$ (GeV) | $p_\pi$ (GeV/$c$) |
|---|---|---|
| $N(1520)$ | 0·61 | 0·74 |
| $N(2650)$ | 3·12 | 3·26 |
| $\Delta(1236)$ | 0·195 | 0·304 |
| $\Delta(2850)$ | 3·71 | 3·85 |

4. The probability of a transition from a state of $\pi^- p$ to some state $f$ can be expressed in terms of a probability amplitude as

$$|\langle f|M|\pi^- p\rangle|^2$$

The total cross-section is obtained by summing over all possible states $f$,

$$\sigma_{\text{tot}}^- = \sum_f |\langle f|M|\pi^- p\rangle|^2$$

$$= \sum_f \langle \pi^- p|M^*|f\rangle\langle f|M|\pi^- p\rangle$$

$$= \langle \pi^- p|X|\pi^- p\rangle \qquad (1)$$

$$|\pi^- p\rangle = |I_3(\pi) = -1, I_3(N) = +\tfrac{1}{2}\rangle$$

$$= \langle 1\tfrac{3}{2}\tfrac{3}{2} -\tfrac{1}{2}|1 -1\tfrac{1}{2} +\tfrac{1}{2}\rangle|I = \tfrac{3}{2}, I_3 = -\tfrac{1}{2}\rangle$$

$$+ \langle 1\tfrac{1}{2}\tfrac{1}{2} -\tfrac{1}{2}|1 -1\tfrac{1}{2} +\tfrac{1}{2}\rangle|I = \tfrac{1}{2}, I_3 = -\tfrac{1}{2}\rangle \qquad (2)$$

where

$$\langle I(\pi), I(N), I, I_3|I(\pi), I_3(\pi), I(N), I_3(N)\rangle$$

are Clebsch–Gordan coefficients. Writing

$$|I = \tfrac{3}{2}, I_3 = -\tfrac{1}{2}\rangle \equiv |\tfrac{3}{2}\rangle$$

$$|I = \tfrac{1}{2}, I_3 = -\tfrac{1}{2}\rangle \equiv |\tfrac{1}{2}\rangle$$

and substituting the values of the Clebsch–Gordan coefficients

$$|\pi^- p\rangle = \sqrt{\tfrac{1}{3}}|\tfrac{3}{2}\rangle - \sqrt{\tfrac{2}{3}}|\tfrac{1}{2}\rangle \qquad (3)$$

Then, since the pion–nucleon interaction conserves isospin,

$$\langle\tfrac{3}{2}|X|\tfrac{1}{2}\rangle = 0 \qquad (4)$$

and

$$\sigma_{tot}^{-} = \tfrac{1}{3}\langle\tfrac{3}{2}|X|\tfrac{3}{2}\rangle + \tfrac{2}{3}\langle\tfrac{1}{2}|X|\tfrac{1}{2}\rangle$$
$$= \tfrac{1}{3}\sigma_{\frac{3}{2}} + \tfrac{2}{3}\sigma_{\frac{1}{2}} \tag{5}$$

Since

$$|\pi^{+}p\rangle = |I_3(\pi) = +1, I_3(N) = +\tfrac{1}{2}\rangle \tag{6}$$
$$= |I = \tfrac{3}{2}, I_3 = \tfrac{3}{2}\rangle$$

$$\sigma_{tot}^{+} = \sigma_{\frac{3}{2}} \tag{7}$$

From equations (5) and (7),

$$\sigma_{\frac{1}{2}} = \tfrac{3}{2}\sigma^{-} - \tfrac{1}{2}\sigma^{+}$$

# Index

absolute conservation laws, 32–3
accelerators, 10, 50, 76–7, 148, 221–5
alpha particles, 4
angular momentum, 6, 74
  addition of, 209–11
  conservation of, 6, 74, 81
  in quantum mechanics, 207–11
antiparticles, 9, 11, 49, 54–7, 72, 141
antiproton, 10, 25, 49
associated production; *see* strange
  particles

*B*; *see* baryon number
baryon, 33, 51
  conservation, 33, 57, 73–4
  decuplet, 112, 134–6, 138–40, 150–151, 172
  multiplets, 133–6, 140–1, 149–51, 172
  number, 33, 51, 53–4, 73–4
  octet, 133–5, 138–40, 150, 154–60, 172
  quark model of, 149–51, 154–5
  Regge trajectories, 166–70
  resonances, 91–114, 125–7
  singlets, 140
beta decay, 17–20, 73–4
  *CP* invariance, 70–2
  non-conservation of parity, 60–6
bosons, 2, 22, 55
Breit-Wigner shape, 94–6, 109–10, 219
bubble chamber, 113–14, 226–9

*C*; *see* charge conjugation
cascade particle, 55
  *see also* Ξ hyperon
centre-of-mass frame, 196–203

Cerenkov counters, 225
channels, 124–6, 214
charge, 40, 53
  conservation, 32, 73–4, 82
charge conjugation, 70–1, 74, 85, 128
Clebsch-Gordan coefficients, 210–11
cloud chamber, 50–3, 148, 226
colliding beams, 191, 224
commutation relations, 35, 207–8
Compton effect, 24
conservation laws, 5, 6, 18, 32–3, 72–4, 81–2
  absolute, 32–3
  angular momentum, 6, 74, 81
  baryon number, 33, 57, 73–4
  charge, 32, 73–4, 82
  energy, 6, 74, 81
  fermions, 82
  hypercharge, 57
  isospin, 39–41, 72–4, 82–3
  lepton number, 79–82
  momentum, 6, 74, 81
  parity, 28, 59, 74, 82–3
  strangeness, 52–4, 57, 74, 82–3
  universal, 81–2
counters; *see* Cerenkov counters,
  scintillation counters
*CP* invariance, 70–2, 74, 82, 85–8
  violation of, 88–92
*CPT* theorem, 70
cross-section, 215
  definition of, 213

Dalitz plots, 107–11, 124
de Broglie wave, 7, 204
decay,
  *K* meson, 50–4, 59–61, 82, 84–92
  Λ, 52–3, 56, 67–8, 83, 247

delay—*contd.*
  muon, 61, 80
  neutron, 18
  Ω, 113–14
  pion, 22–3, 32, 60–1, 76–8
  Σ, 73, 82
  Ξ, 55
  *see also* beta decay
decuplets, 112, 134–5, 138–40, 172
delta function, 187
Δ resonances, 95–103, 112, 134, 140
detailed balance, principle of, 26, 214–16
detectors, 225–31
deuteron, 26, 30, 39–41
dilatation (of time), 203
Dirac theory of relativistic fermions, 8, 45

*e*; *see* electron
effective mass, 109
eigenfunctions, 206–7
eigenstates, 207
eightfold way, 136
  *see also* $SU(3)$
electric charge, *see* charge
electromagnetic interactions; *see* interactions, electromagnetic
electromagnetic force, 21
electron, 2, 40, 49, 76, 78–80
  beta decay, 17–20, 62–6
  -electron scattering, 17
  magnetic moment, 45–8
  Mott scattering, 178
  -neutron scattering, 182–3
  polarization in beta decay, 62–6
  -positron pair production and annihilation, 9–14
  -positron scattering, 17, 191
  -proton scattering, 14–15, 178–81, 183–91
  scattering, 63–5, 175–91
energy,
  conservation of, 6, 74, 81
η meson, 118–19, 127, 129, 141–3
exchange forces, 165
exclusion principle, *see* Pauli exclusion principle

fermions, 3
  conservation of, 82
Feynman diagrams, 11–17, 21, 24
  annihilation, 13
  beta decay, 19, 80
  Compton effect, 24
  electron scattering, 14, 15, 17
  muon decay, 80
  nucleon, 44–5
  pair production, 13, 14, 24
  pion decay, 81
  weak interaction, 19, 80–1
form factor, 176–83
four momentum, 178, 184, 195
four vector, 186, 195–6

$(g - 2)$ experiment, 46–8
$G$ parity, 128–9
$\gamma$; *see* photon
Gell-Mann–Okubo mass formula, 138–42, 154–60
gravitational interaction, 73
group, 136–8

hadron, 126
half life, 212
handedness, 35, 66–7, 70–2
Heisenberg uncertainty principle 14, 21, 93, 218–19
hermitean adjoint, 206
hermitean operator, 206
hypercharge, 56–7, 126–7
hyperons, 51
  *see also* Λ-, Ω-, Σ-, Ξ- hyperon

$I$; *see* isospin
interactions,
  electromagnetic, 13–17, 39, 40, 45–6, 72–4, 173–4, 175–91
  gravitational, 73
  strong, 52–4, 56, 72–4, 93, 126
  superstrong, 134
  types of, 72–4
  weak, 52, 60, 67–9, 73–4, 76–83, 113
intersecting storage rings; *see* storage rings
invariance, 5, 6
  charge conjugation, 70–1, 74
  $CP$, 70–2, 74, 82, 85–8

inversion; *see* parity
isospin, 39–41, 72–4, 82–3
*PCT*, 70, 74
reflection; *see* parity
time reversal, 69, 70, 74
invariant mass, 108
inverse beta decay, 20, 78
inversion, 27, 69
    effect on spin, 61–2
    *see also* parity
isobaric spin; *see* isospin
isospin, 35–42, 53–7, 72–4, 82–3, 95,
    126–8, 131–3, 136, 156, 249
isotopic spin; *see* isospin
*I* spin; *see* isospin

Jacobean peak, 120–3, 202

*K* meson, 51–61, 67–8, 84–92, 141–3
    decay, 50–4, 59–61, 82, 84–92
    $K_L$, $K_S$, 91–2
    $K_1$, $K_2$, 85–91
    neutral, 54, 84–92
    non-conservation of *CP*, 88–92
    scattering, 110–11
*K* resonances, 123–4, 127, 141–3

laboratory frame, 196–203
$\Lambda$ hyperon, 51–8, 73, 133–5, 154
    decay, 52–3, 56, 67–8, 83, 247
$\Lambda$ resonances, 110–11, 125–7, 139–
    140
lepton,
    conservation, 79, 81–2
    definition of, 79
    number, 79–80
    two kinds of, 79
life-time, 212, 218–20
Lorentz transformation, 194–5

magnetic moment,
    electron, 45–8
    muon, 46–8
    nucleons, 43–5, 173–4
magneton,
    Bohr, 241
    nuclear, 43, 242
mass energy, 1, 2
mass formula; *see* Gell-Mann-
    Okubo mass formula

mean lifetime, 212
meson, 20–2, 56–7
    nonets, 142–4, 146, 152–4
    quark model of, 146–7, 151–4
    Regge trajectories, 167–71
    resonances, 114–24, 128–9, 141–4
    *see also* $\eta$, muon, $\omega$, $\varphi$, pion, $\rho$
missing-mass spectrometer, 119–23,
    202
momentum, 1
    conservation of, 6, 74, 81
Mott scattering, 178
$\mu$-meson; *see* muon
multiplets, 35, 39, 133–60
muon, 21–3, 40, 49, 76–83
    decay, 61, 80
    magnetic moment, 46–8

*N*; *see* nucleon, *N* resonances
*n*; *see* neutron
*N* resonances, 95–103, 139–40
    *see also* $\Delta$ resonances
negative-energy states, 8–10
neutrino, 17–20, 49, 76–83
    electron-, 17–20, 76, 79, 80, 82
    muon-, 76–83
    two-component, 66–7, 70–2
    two kinds of, 76–8
neutron, 4, 35–45, 49, 56–7, 133, 135
    decay of, 18
    form factors, 182–3
    magnetic moment, 43–5, 173–4
nonets, 142–4
nuclear force, 20–1, 38–9
nuclear magneton, 43, 242
nucleon, 35–9, 41–3, 133
    definition, 35
    electromagnetic structure, 178–
    191
    magnetic moment, 43, 173–4
    parton model of, 186–91

octets, 134–5, 138–44, 146–7, 172
$\Omega$ hyperon, 112–14, 134–6, 139
    decay, 113–14
$\omega$ meson, 95, 115–19
operators, 205–7

*P*, 27
    *see* parity

*p*; *see* proton
parity, 26–30, 32–3, 59–71, 74, 137
  conservation of, 28, 74, 82–3
  intrinsic, 29, 84–5
  non-conservation of, 59–68, 70–4
  of *K* mesons, 84–6
  of negative pion, 30, 34
  of neutral pion, 32
  of neutron, 30, 32–4
  of positive pion, 32
  of proton, 30, 32–4
partial wave cross-sections, 98, 103
Pauli exclusion principle, 3, 4
*PC*; *see CP*
phase space, 104, 107, 115, 214
$\varphi$ meson, 129, 141
photon, 1, 9, 13–16, 175
$\pi$ meson; *see* pion
pion, 21–34, 39–45, 56–7, 59–60, 76,
    127, 129, 141–3
  charged, parity, 30
  charged, spin, 26
  decay, 22–3, 32, 76–8, 81
  neutral, spin and parity, 31
  -nucleon scattering, 95–103
  resonances, 114–23
  Yukawa theory, 20–2
positron, 8–14, 17, 191
  beta decay, 17–20
proton, 3, 35–45, 56–7, 133–5
  form factors, 178–81
  magnetic moment, 32–5, 173–4
pseudoscalar, 29, 31, 141–2

*Q*; *see* charge
quantum mechanics, 204–11
quarks, 144–55, 172–4
  properties, 147–8

reflections, 27–8
Regge poles, 162–71
relativity, 1, 58, 104, 108–9, 120,
    178, 186–7, 194–203
resonance, 93–132
  of classical oscillator, 217–20
$\rho$ meson 114–15, 118–22, 129
Rutherford scattering, 177

*S*; *see* strangeness

scattering,
  cross-section, 213, 215
  differential cross-section, 197–8,
    213
  electron, 63–5, 175–91
  electron-electron, 17
  electron-neutron, 182–3
  electron-positron, 17, 191
  electron-proton, 14–15, 178–81,
    183–91
  form factor, 176–83
  *K* meson, 110–11
  Mott, 178
  partial wave cross-sections, 98,
    103
  perturbation theory, 176
  pion-nucleon, 95–103
  relation in laboratory and centre-
    of-mass frames, 197–203
  resonance, 93–4
  Rutherford, 177
Schrödinger equation, 204–5
scintillation counters, 20, 225
$\Sigma$ hyperon, 51, 55–8, 73, 134–5, 154
  decay, 73, 82
$\Sigma$ resonances, 106–9, 112, 126–7,
    135, 139–40
signature, 166, 168
spark chamber, 76–8, 229–31
spin,
  of electron, 3
  of neutral pion, 31–2
  of neutrino, 18
  of neutron, 5
  of positive pion, 26
  of photon, 2
  of proton, 3
  unaltered by inversion, 61–2
statistics, 2, 3
state, quantum mechanical, 205–6
storage rings, 222–5
strange particles, 49–58
  associated production, 51–3
  discovery of, 50–1
  *see also* hypercharge, strangeness
strangeness, 52–7, 73–4, 82–3, 126–7
strong interactions; *see* interactions,
    strong
*SU*(3), 136, 138–47, 149–60, 172
*SU*(6), 172–74

symmetry, 74
  *see also* group, invariance, $SU(3)$,
    $SU(6)$
synchrotron; *see* accelerator

$T$; *see* time reversal
$\tau$, 59–61
$\theta$, 59–61
time dilatation, 203
time-reversal, 69–70, 74
trajectory, Regge, 162–71
two-component neutrino, 66–7, 70–2
two neutrinos, 76–7

uncertainty principle; *see* Heisenberg uncertainty principle
unimodular matrix, 136
unitary matrix, 136
unitary symmetry; *see* $SU(3)$
units, 241–2

universal conservation laws, 32–3,
  81–2
$U$-spin, 156–9

$V$-particles, 50
virtual processes, 13, 43–6, 81

wave function, 204–5
weak interactions; *see* beta decay;
  interactions, weak
weight diagrams, 146

$\Xi$ hyperon, 51, 55–8, 134–5
  decay, 55
$\Xi$ resonances, 109, 112, 126–7, 135,
  140

$Y$; *see* hypercharge
$Y$ resonances; *see* $\Lambda$ resonances, $\Sigma$
  resonances
Yukawa theory of nuclear forces,
  20–2